PLANT PROPAGATION for the AMATEUR GARDENER

PLANT PROPAGATION
for the
AMATEUR GARDENER

John I. Wright

BLANDFORD PRESS

Poole   Dorset

First published in the U.K. 1983 by Blandford
Press, Link House, West Street, Poole, Dorset,
BH15 1LL

Distributed in the United States by Sterling
Publishing Co., Inc., 2 Park Avenue, New York,
N.Y. 10016.

British Library Cataloguing in Publication Data

Wright, John I.
   Plant propagation for the amateur gardener.
   1. Plant propagation
   I. Title
   635'.043   SB119

ISBN 0 7137 1155 8

Typeset by Polyglot Pte Ltd, Singapore

Printed by Toppan Printing Co. (S) Pte. Ltd.

Contents

Part 1
METHODS of PROPAGATION

1 Cuttings

Raising plants from cuttings is a quick, easy and cheap method of increasing the plants in your garden or greenhouse. Many kinds can be propagated in this way, and cuttings are so easy to take that any visit to a friend's garden can yield a few new specimens with no detrimental effect on either the parent plants or your friendship.

A cutting is any part of a plant which, when taken from its parent and treated in the correct manner, becomes an individual plant in its own right. Most cuttings are stem cuttings, but in some cases plants can also be propagated from their leaves and roots.

As the rooted cutting will be identical to its parent, great care should be taken when selecting that parent. A poor-growing, spindly plant is unlikely to produce a prize specimen from a cutting, and neither is a plant riddled with disease. Plants selected for cutting propagation should therefore be the strongest-growing, most healthy plants available.

Cuttings of some type can be taken throughout the year, the owner of a heated greenhouse quite naturally having the greatest choice of subject and propagating season; yet even without such facilities there is rarely a time of the year when something cannot be prepared as a cutting.

Almost all cuttings are prepared in the same way before planting. They should all be taken from the parent plant with a sharp knife or razorblade. Sharpness is essential as bruising caused by a blunt knife will possibly result in the cut end of the cutting rotting away and subsequently failing. For this same reason, cuttings should never be taken or prepared with the fingernails; although they may be broken off by bending in some cases, a knife only being used if any trimming of loose ends seems necessary.

Most stem cuttings should be taken immediately below a leaf-joint or node (*see* 'Internodal and Nodal Cuttings'), making the cut as straight as possible across the stem. Once removed from the parent plant any leaves, stalks or buds likely to be beneath the surface of the soil when the cutting is planted should be taken from as close to the stem as possible with the knife. When the cutting has been prepared in this manner it is ready for planting (Fig. 1).

To assist with root formation, the cutting is best dipped into a hormone rooting compound of a suitable type, first dipping it into a little clean water to assist the powder to stick to the stem. Although some people disagree, it has been my experience that saving the powders from one year to the next is false economy, as the effectiveness appears to wear off with exposure to the air. For my part I buy fresh stock in each season. Cuttings should be dipped into the powder to a depth in relation to their size; 6 mm or so for the smallest softwood to perhaps 4 cm for tall hardwood cuttings.

Leaf and root cuttings are prepared in a different manner, of course, and this is described later, but whichever method is used for taking or preparing a cutting a correctly formulated compost is of prime importance for the successful rooting of a large number of subjects.

There are a number of formulae for making up composts suitable for rooting cuttings, or 'striking' them, as the process is generally known. For most purposes the John Innes Seed Compost is ideal. This, along with proprietary seed composts, can be bought from garden centres etc, or it can be made up yourself from the separate ingredients.

The measurements of these various components is by volume, the basic unit being the bushel (36 dry litres or 8 dry gallons), or for larger quantities, the cubic metre (cubic yard). As most gardener's requirements are likely to be in bushels rather than cubic metres it is convenient to have a container in which the former amount can be measured. Four 2-gallon buckets yield a bushel, but a rectangular wooden box specially made for the purpose is best. A box with an internal measurement of $56 \times 25 \times 25$ cm will hold 1 bushel of soil or compost. The following are some of the compost formulae suitable for cuttings.

John Innes Seed Compost
2 parts (by bulk) sterilised loam.
1 part moss peat.
1 part coarse sand or grit.
To each bushel is added 40 g 18% Superphosphate, 20 g chalk.

Soilless compost
You can buy peat-based proprietary composts, but if you use one of these choose a seed and cutting, or a 'universal', formula. It is easy to make your own soilless cutting compost, however, from silver sand (2 parts by bulk) and moss peat (1 part). As this mixture contains no nutrients, the rooted cuttings will have to be potted up without delay.

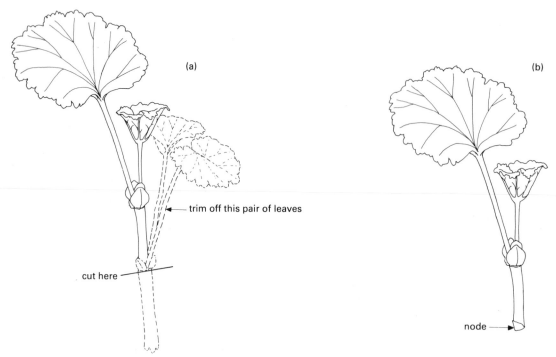

trim off this pair of leaves

cut here

node

Fig. 1 (a) Softwood cutting from *Pelargonium* (geranium). (b) Cutting prepared for planting.

Various other materials can be used as the medium for raising cuttings, the most popular probably being the substance vermiculite. This is a type of mica found in the USA and South Africa which, when heated and expanded, becomes light and flaky and has a great capacity for retaining water without becoming waterlogged. As with other soilless composts it is essential to transplant or feed cuttings rooted in such material as soon as roots are formed. The vermiculite used for loft insulation may be too alkaline, so it is wise to choose 'horticultural vermiculite', which will have had the pH adjusted.

There is no special requirement as to the type of container suitable to hold the compost in which to root cuttings, the only real essential being that it has free drainage. Although wooden boxes, plastic trays and pots are all suitable it will be found that many cuttings root especially well if planted around the rim of a clay pot.

Watering cuttings can often prove a difficulty, as too much water can cause the cutting to rot below ground and too little will cause drying out. The best rule to follow is to start off with moist compost and give the cuttings a good watering with a fine rose immediately after planting, then leave them alone. If the weather is hot then spray them over lightly night and morning but do resist the temptation to overwater. Remember that many softwood cuttings will droop for the first few days after planting in any case, this being a natural reaction to being separated from the roots of the parent, and further watering would not have any effect on this.

Once cuttings are rooted they are transplanted, more often than not into individual 7.5 cm pots, or they may be planted out in boxes or in frames. Those transplanted into pots or boxes should be treated in a similar manner to seedlings, and details of composts can

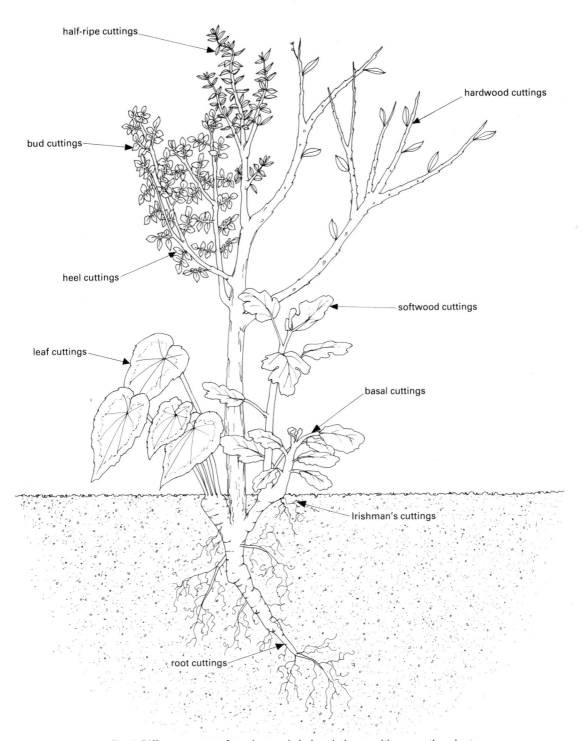

half-ripe cuttings

hardwood cuttings

bud cuttings

heel cuttings

softwood cuttings

leaf cuttings

basal cuttings

Irishman's cuttings

root cuttings

Fig. 2 Different types of cutting and their relative positions on the plant.

be found in Chapter 2. More information is given under the individual plant names.

There are several different types of cuttings (Fig. 2), which for ease I will describe under their appropriate names.

BASAL CUTTINGS

A basal cutting is a young shoot cut or gently pulled from the base of a parent plant. Generally, such cuttings need little preparation other than the removal of a few lower leaves before planting. They should be taken when they are as young and soft as is convenient to handle. A cutting length 2.5–10 cm is most usual. They need some form of protection to avoid drying out while rooting, the heated propagating frame being best, especially early on in the season when most cuttings of this type are taken.

Many perennials and alpine plants can be propagated by means of basal cuttings.

HEEL CUTTINGS

A heel cutting is usually a sideshoot of hardwood or semi-hardwood pulled from the parent plant complete with a 'heel' of bark from the main stem (Fig. 3). Many shrubs such as *Jasminum nudiflorum* are found to root much better from such cuttings. They should be prepared and grown as ordinary hardwood and semi-hardwood subjects, but without doing anything to the heel at the base of the cutting other than trimming any loose ends with a sharp knife before planting.

Although perhaps more a basal than a heel cutting, it will be found that a dahlia cutting will root more readily if pulled or cut from the crown of the tuber with a little bark or rind attached. Certainly, the speed of rooting is much better in cuttings treated in this way.

BUD CUTTINGS

Taking a bud cutting (Fig. 3) is very similar to preparing a bud for budding, the main difference being that with the cutting the leaf is retained attached to the slice of stem.

A semi-hardwood shoot should be chosen to supply the bud, usually in the late summer from the same season's growth. A slice of stem around 2.5 cm in length is cut from the parent plant, each cutting containing a leaf and a leaf axil containing a bud. Plant the cutting so that the bud and leaf are just above the surface of the

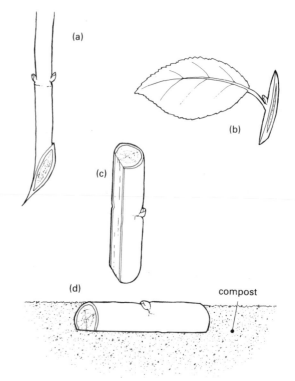

Fig. 3 (a) Heel cutting. (b) Bud cutting. (c) Eye cutting. (d) An eye cutting pressed into compost.

compost and place in a frame or propagator with gentle bottom heat until well rooted.

Camellias and roses are among the plants which can be propagated by this method.

EYE CUTTINGS

This is a very successful method of propagation usually used for vines. An eye cutting is a hardwood stem cutting taken from the parent plant during the dormant periods of autumn and winter.

The best cutting is one about 4 cm long, containing one 'eye' or bud (Fig. 3). A length of vine stem can be cut into many sections each containing a bud. With the sections so cut, a strip of bark is removed from the opposite side of the stem to the bud, and the cutting pegged down on a loamy compost, J.I. Seed Compost being ideal.

With the cutting lightly pegged down, sprinkle silver sand liberally over the compost surface and press the cutting in so that only the bud protrudes above the sand. A bottom heat of 24 °C is required for rooting, making a propagating frame of some kind essential.

With this temperature the compost is inclined to dry out rapidly, so it is a good plan to check periodically to make sure that it remains moist around the cutting.

STEM CUTTINGS

Most cuttings fall into this category. A stem cutting is any cutting taken from the main shoot of a plant or any sideshoots growing from the same plant or stem. The category is divided into three sections; softwood, semi-hardwood and hardwood cuttings. Particularly with shrubs and trees, cuttings from young plants are to be preferred to cuttings from very old ones as for some reason they seem to produce roots more easily.

Softwood cuttings
This is a group name given to any cutting taken from shoots that are still soft and have not become hard or 'woody'. Usually, the cutting is 4–7.5 cm long, and is prepared like other stem cuttings with the lower leaves removed and the stem cut immediately below a node with a razorblade or knife before planting.

As softwood cuttings are by nature soft and fleshy, they are easily dried out and killed, so they have to be rooted in a close damp atmosphere and need at least a little bottom heat to encourage root formation. A propagating frame with a temperature controlled at 16–18 °C is ideal.

A sandy type of compost suits them best, and when planting it is a good rule not to plant too deeply; just enough to hold the new plant upright is sufficient (Fig. 4).

Many herbaceous plants as well as chrysanthemums, dahlias and geraniums are propagated from softwood cuttings.

Half-ripe or semi-hardwood cuttings
The term 'semi-hardwood' covers a number of types of cutting, but basically it means that the cutting is prepared from wood of the current season's growth. This is usually found on the sideshoots or on the top of the main stem of the parent plant in late summer.

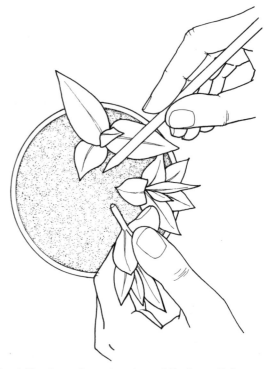

Fig. 4 Planting softwood cuttings of *Tradescantia* in a small pot.

Cuttings are usually 5 to 10 cm long, and are prepared by trimming the cutting with a straight cut below a node and removing a few lower leaves. Planting is best done in J.I. Seed Compost or a peat/sand mixture, the cuttings being inserted to a quarter of their length. A shaded cold frame is the best situation for rooting semi-hardwoods. Very many species of shrub and shrubby plants can be propagated from cuttings of this type (Fig. 5).

Hardwood cuttings
This type of cutting is ideal for many hardy shrubs, trees and soft fruits, and is one of the easiest to handle. The hardwood cutting is a stem cutting taken from the parent plant at the end of the growing season when the wood of the current season's growth has matured ready for the winter. October and November are the months when most of such cuttings are taken.

.

Cuttings of 25–30 cm should be cut from the parent plant just below a node, and any lower leaves still attached should be removed. Any well-drained and slightly shaded corner of the open garden will be suitable to root these cuttings. Planted out, they will remain dormant throughout the winter, breaking into growth the following spring. In colder districts it is better to plant them in the soil of a cold frame to give them extra protection, and this is what I would recommend.

Planting any quantity is easily done by making a V-shaped trench 10 cm or so deep with a spade. This is done by simply pushing the spade into the soil and waggling it back and forth a couple of times. In anything other than very light soils a small amount of silver sand is best spread along the bottom of the V to assist with drainage and encourage rooting. Place the cuttings 15 cm apart along one side of the trench and simply press the soil back against them with the side of the foot. Single cuttings are easily planted with a dibber.

After severe winter weather it may be found that the frost has lifted or loosened the cuttings in the ground,

and any affected in this way should be firmed back into position with the foot. Plants rooted this way are generally ready for planting in their final positions in the following autumn.

TIP CUTTINGS

This is a general term, covering all those cuttings which are taken from the growing tip of a non-flowering shoot. They should be treated in the same way as softwood cuttings.

INTERNODAL AND NODAL CUTTINGS

The node of a plant stem is that part of the plant where the leaf-stalk or petiole joins with the stem. Most cuttings root best if they are cut across in a straight line immediately below a node, and are therefore termed 'nodal' cuttings. This benefit to rooting is especially

Fig. 5 (a) A half-ripe nodal cutting of *Weigela* taken in August. (b) *Weigela* cutting rooted and ready for potting on in March after overwintering in a cold frame.

noticeable in plants inclined to have a hollow stem, as the node generally leads to a branch of solid tissue throughout such stems.

Certain plants though appear to root better if the stem is cut midway between a pair of nodes. These cuttings are therefore termed 'internodal'. Clematis is one plant propagated from this type of cutting (Fig. 6).

IRISHMAN'S CUTTINGS

An Irishman's cutting is a basal cutting which is pulled from the crown of a parent plant complete with a few roots already attached. It is perhaps really a mild form of division.

Early-flowering chrysanthemums and other herbaceous perennials can quite often be propagated in this way. The rooted cutting is potted on with little need to spend any time in the propagating frame, although I would recommend a brief stay until the root system is well established.

LEAF CUTTINGS

A wide range of greenhouse plants can be propagated by means of various types of leaf cutting. Usually, a complete leaf, well developed and with its stalk attached, is removed from the parent plant. The leaf stalk is then pushed into a compost of sand and peat until the leaf itself is lying flat against the surface. In a short time roots are developed at the base of the leaf and a new plant is formed. Gloxinias and Saintpaulias are two of several plants propagated in this way.

A slightly different treatment is given to the leaves of *Begonia rex*. A crisp mature leaf is taken from the parent plant with merely a stub of leaf-stalk, and a number of cuts 2.5 cm or so apart are made in the well-defined main veins on the back of the leaf (Fig. 7).

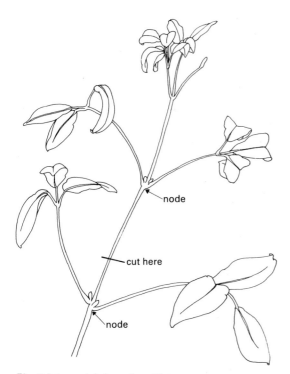

Fig. 6 Internodal clematis cutting.

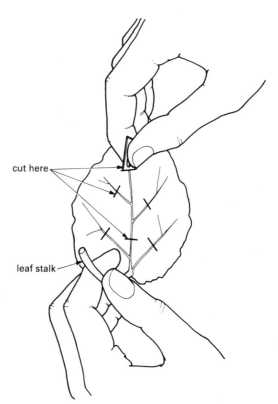

Fig. 7 Cutting the main leaf veins of the underside of a leaf cutting of *Begonia rex*.

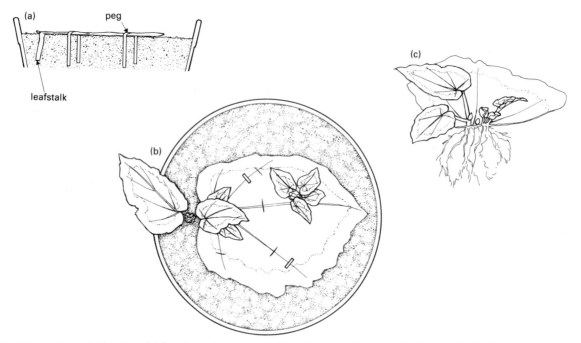

(a)

peg

leafstalk

(b)

(c)

Fig. 8 Begonia rex leaf cutting. (a) Peg the leaf down so that the severed veins are in close contact with the compost. (b) New plants form on the cut veins. (c) A young plant separated and ready for potting on.

With these cuts made, the leaf is pegged down on a peaty compost so that the cuts come into close contact with the medium. Eventually roots will form at each cut and from them new plants will develop (Fig. 8). Such leaves can also be cut into about 2.5 cm pieces, pegging each one to the compost in the way described. In this way many new individuals can be raised from a single leaf.

Sansevierias (Mother-in-law's Tongue) are propagated in a similar fashion, only this time the leaves are cut across to form sections about 5 cm long. These are planted upright in a sandy compost in the same way as ordinary softwood cuttings.

Leaf cuttings need to be rooted in a close, damp atmosphere; so it is important to keep everything used in the propagating procedure as clean as possible, compost and containers being sterilised if necessary. It is also important to achieve the right balance with watering: too much and the leaves will rot, too little and they will shrivel up. As a general guide to the dampness required take a handful of peat compost and squeeze it tightly in the hand. It should be possible to wring a few drops of water from it if the moisture content is correct.

Most subjects will require a temperature of between 16 °C and 18 °C to root successfully, so making use of a heated propagator or similar a virtual necessity. The atmosphere in the propagator should be kept damp by a daily syringe with warm water.

PIPING CUTTINGS

A piping cutting is simply a stem cutting taken from the tip of a carnation or pink. It is very easy to take, the tip of any non-flowering shoot being suitable.

Hold the selected stem with one hand while pulling the tip carefully with the other, taking care that the cutting parts from its parent just above a node. No other treatment is then necessary, the piping simply being dipped into a rooting compound and dibbled into a sandy compost to root. A cold frame is usually sufficient protection for these cuttings.

ROOT CUTTINGS

Perhaps rather surprisingly, a large number of plants can be propagated from cuttings taken from their roots (Fig. 9), ranging from the Californian tree poppy (*Romneya coulteri*) to the humble horseradish. The method used for all is much the same.

Cuttings are generally taken during the winter months when the plants are dormant. Sections of thicker roots are cut into pieces 5–7.5 cm long, the cut nearest the crown of the plant being made straight across and the lower cut being on a slant, so that you will know which way to insert them. The cuttings are inserted into a suitable compost, such as J.I. Seed Compost, leaving the top and straight end of the cutting just below the surface. A cold frame or greenhouse is all that is necessary for the protection and propagation of the hardy species. By spring the cutting will have branched out and buds will appear at the crown.

A different technique has to be used for thin-rooted subjects, such as phlox. With this type of root it is best to cut them into sections as before but lay them flat on the surface of a box partly filled with a suitable compost, covering them with a shallow layer of the same material. As with the thicker roots, the new shoots will appear in the spring. Plants produced in this way are transplanted into nursery beds in the garden in the following summer.

Certain plants, those of the genus *Acanthus* for example, have a juvenile type of foliage during their early stages of growth. Root cuttings taken from such plants, however, tend to take on the maturity of their parent; therefore, cuttings from a mature plant of this type will develop mature foliage whereas those from a young plant will develop juvenile foliage. As a degree of maturity is required before flowering can take place, it follows that root cuttings from a mature plant are to be preferred.

Another thing to note is that root cuttings from variegated hybrid plants often revert to the green of the original plant, and so should be avoided.

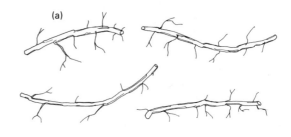

(a)

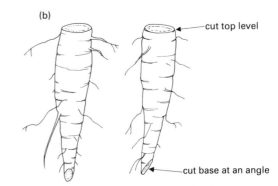

(b)

cut top level

cut base at an angle

Fig. 9 Root cuttings. (a) Perennial *Phlox*. Such thin types of root are laid horizontally in the compost to grow. (b) Thicker roots such as horseradish should be planted vertically with their tops level with the compost surface.

2 Seeds

Without doubt there are more plants raised for our gardens from seed than by any other method. Annuals and vegetables are almost exclusively propagated in this way.

There are three main advantages in raising plants from seed. Firstly, seed is a relatively cheap way of obtaining large numbers of plants. Secondly, good seed has less chance of carrying disease than vegetatively-produced plants, and, thirdly, plants raised in the environment in which they will mature will be stronger and more tolerant than their imported cousins.

All seeds, no matter what their type, require the same basic elements to effect their germination. These are moisture, air, light and heat. It is on the degree and manner in which these are given that success or failure depends.

Vegetables and the hardy annuals probably represent the easiest of all seeds to raise. Both types have been bred or selected to give the best results under variable conditions. As long as the basic rules are followed they have a wide degree of tolerance.

It follows that the more exotic the species the more exacting will be its requirements, and in all aspects of cultivation it helps to understand the plant and the environment it inhabits in its natural state. Obviously, the alpine from the high ranges of the Himalayas requires a different set of conditions for germination than a tender plant from the steaming jungles of Brazil.

TYPES OF SEEDS AND METHODS OF ENCOURAGING GERMINATION

Seeds come in all shapes and sizes, from the minute, almost microscopic, seeds of the orchids to the massive 18 kg seeds of the double coconut; but most types fall into recognisable groups bearing the same characteristics. There are five main natural groups: dust-like seeds, hard-coated seeds, fleshy seeds, oily seeds and winged or plumed seeds. There are also the 'pelleted' seeds (Fig. 10).

Pelleted seeds are usually very small seeds which are coated in an inert substance to make them larger

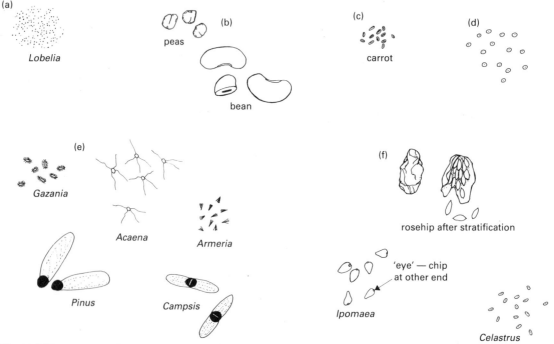

Fig. 10 Different types of seed: (a) dustlike; (b) fleshy; (c) oily; (d) pelleted; (e) winged and plumed; (f) hard-coated.

and/or a more even shape. This enables you to sow them more easily, either by machine or hand. The seeds can be set out at the correct planting distance from the start. As a general rule, pelleted seed should be sown less deep than would be normal for a given seed and at all times the soil or compost used should be kept in an evenly moist condition until germination takes place.

The various types of seeds have differing properties and requirements. *Dust-like seeds*, such as those of begonias, generally do not have a long storage life. They are invariably sown on the surface of the compost and are merely pressed in.

Because their seeds are impervious to water, *hard-coated types* are often a problem as they will not germinate until this protective covering has been broken down. This is achieved in various ways before sowing.

Sweet pea varieties prone to this trait are often 'chipped', which means that a small sliver of the skin is removed from the seed opposite the eye. Care should be taken not to damage the seed with too enthusiastic a cut, and it is perhaps safer to use a small file to wear away a portion of the surface.

Freshly gathered hard-coated and flesh-covered seed such as the holly produce better results if they are 'stratified'. This is done by placing the seeds in layers in a pot or box containing a mixture of sand and peat in equal quantities. These are then placed outside where frost and snow can work on the seed coat through the winter and soften it. Mice and birds are the enemy here and containers are best protected with a covering of fine wire netting.

Many alpine seeds require this type of cold treatment. Normally it can be achieved by sowing early in the year and exposing the pots to weather throughout February and March. This can be artificially reproduced by placing the seeds, either in their moistened seed packet or preferably already sown in a moist compost, leaving them overnight to absorb the moisture before placing them in the freezer compartment of a fridge for a time. Forty-eight hours is said to be enough, but I always leave mine in for around a week. It is most surprisingly effective.

Quite the opposite of cold is the treatment required by some eucalypts, brooms and other heath/forest plants which only germinate freely after being subjected to fire and heat in some degree. Such seeds can be given this treatment by placing them in a metal sieve and passing them over a gentle flame until a few burst, indicating that they have reached the required temperature. When this

happens the remainder are plunged immediately into a container of slightly warm water for a few moments before sowing in the normal way.

Fleshy seed is the term used to describe such seeds as peas and broad beans. These respond well to being soaked in clean water for about 24 hours before sowing. Incidentally, this method will also determine whether a variety of sweet pea is hard-coated or not—those seeds which require chipping remain afloat after their more receptive cousins have absorbed moisture and sunk.

There is little difficulty with the germination of *oily seeds* other than to note that such seeds soon shrivel. Carrot and magnolia are of this type and should be sown as soon as they are ripe, or the season they are received as they will not keep well in storage.

Lastly come those seeds intended for dispersal by the wind. Such seeds have wings as in those of the sycamore, or plumes as in seeds of the dandelion. With *winged seeds* the wings are best removed before sowing.

Temperatures for optimum seed germination are as varied as the seeds themselves, but generally most respond well to a temperature of 5–6 °C higher than the natural temperature in which the plant achieves full growth. In general *very* high temperatures are to be avoided unless specified as they can inhibit germination rather than encourage it, particularly with certain plants, such as lettuces.

Although the seed of highly bred garden varieties is likely to germinate relatively evenly and quickly, the seed of many species is often slow to germinate. Some seed takes two years or more, and with such types it is best not to assume failure until at least this amount of time has passed. During this waiting period it is essential to keep the seed compost moist and to protect the seeds from the attack of mice, birds and insects.

COLLECTING AND PREPARING SEED

Whereas our climate allows us to grow numerous plants to their flowering stage, in many cases it is a rare year when we are graced with a summer and autumn warm and dry enough for their seeds to ripen adequately. This and various other factors make seed-saving by the amateur a rather restricted affair.

Nevertheless, to grow a crop from your own seed gives a great deal of satisfaction and many plants do oblige by setting seed, some more enthusiastically than would be wished.

As seed ripening follows flowering it will be obvious that those flowers and vegetables reaching maturity earlier in the year will be the most likely to set seed and ripen. In the vegetable garden peas, various kinds of beans, lettuces and tomatoes all oblige by maturing early on self-pollinating plants. Biennial vegetables, too, such as beetroot, carrots and onions, are also suitable, being grown on to flowering in their second season after winter storage.

One of the difficulties in the average garden is making sure that plants are not cross-pollinated, which is why those that are self-pollinating are best suited. The cabbage family in particular cross-pollinates easily, and seed saved from any could yield a hotch-potch of cabbage types, mostly useless, and none or very few showing the characteristics of its immediate seed parent. If an attempt is made with these types of plant it is essential that only one variety should be allowed to flower, thus giving the likeliest chance of true seed (and even then F_1 hybrids will not produce offspring like the parent).

Plants selected for seed production should be chosen for their health, vitality and closeness to the 'type' of plant required. With tomatoes, for instance, you could select the plant bearing the tastiest fruit or conversely one bearing the heaviest crop. In an ideal world the two might be combined. When selecting always remember that characteristics given to a plant by its growing environment will not be passed on to its seed.

Peas and beans are perhaps the easiest seed for the beginner to save as they are easily seen and handled. Seeds should be saved from those pods appearing first, thus giving more time for successful ripening. A part of a row can be set aside for the purpose or the lower pods left to ripen while those above are picked.

With all seed-saving ventures it is important to mark the plant or plants selected as seed parents clearly. The 'head gardener' is not always the one who picks the vegetables or flowers and many a prospective seed harvest has ended up in the pot or the table decoration!

Dry weather is essential for good ripening and in the case of peas and beans these are known to be ripe when the pods begin to look dryish and yellow and begin to split (Fig. 11). In wet seasons the plants are best pulled up whole just prior to ripening and hung up in a well ventilated shed or greenhouse where the rain cannot reach them.

When dry the pods are taken singly from the plant as they split and the seed removed, placing them in shallow containers to dry fully in a warm greenhouse, shed or on a sunny windowsill. As with most seeds, when dry they should be stored in unsealed paper bags in a cool, dry and airy place. At all times seed stocks should remain clearly labelled.

The seeds of tomatoes, and of other similar fleshy fruits, are more complicated to prepare for storage, the first essential being that the fruit should be really ripe before it is picked. Tomatoes are very soft, almost mushy to the touch when they are fully ripe. Vegetable marrows on the other hand are hard and sound slightly hollow if lightly tapped with the knuckles. It is difficult to know whether cucumbers are ripe or not, and it is best to leave them hanging on the plant until well into autumn before preparing them.

Commercially, these types of seed are removed from the fruit, first by pulping them (not crushing them!) in a wooden or earthenware container and then treating them with commercial hydrochloric acid at the rate of 1 fluid ounce to 500 g of fruit. This mixture is left for varying periods depending on the plant and separates the seeds from the pulp after which the seeds are washed thoroughly in fresh water.

This method can be used by anyone, of course, but a less costly and safer method is best suited for home use. The ripe fruit is pulped in the same manner into a bottling jar or similar and allowed to ferment in a warmish place for two or three days, with an occasional stir. This will cause the seed to separate from the pulp to a large degree whereupon the whole mess can be placed in a fine sieve and the seed washed clean under running water. After separation and washing, the seed is spread thinly on a clean cloth to dry in a warm place before storing.

Seeds can be saved from the many ornamental garden plants which are true species. Seed from hybrid forms and varieties should be avoided as they are unlikely to come true to type. Such plants are best propagated vegetatively from cuttings etc. Having said this, however, it must be remembered that most hybrids have been raised from seed originally, and experimenting can be fun if you have the time and space.

Nevertheless, if we leave aside the various hybrid types of plants in our gardens, a quick inventory will reveal many which are true species, and amongst these are many shrubs, trees, alpines and border plants.

One of the advantages of collecting your own seed is that it can be sown as soon as it is ripe, often resulting in much improved germination. Primulas, for example, are notorious for slow and haphazard germination after

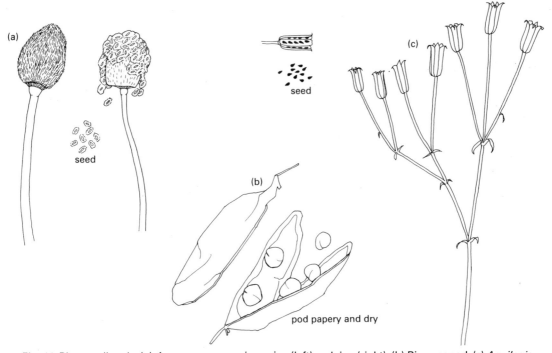

Fig. 11 Ripe seedheads. (a) *Anemone coronaria*, unripe (left) and ripe (right). (b) Ripe peapod. (c) *Aquilegia*.

storage, yet fresh seed germinates easily. With all self-saved seed of species plants it is a good plan to sow a portion of it as soon as it is collected and store the remainder for sowing in the spring.

Keen observation and dry weather are the two real necessities for successful seed collection. A careful watch should be kept on the ripening seed heads so that they can be plucked from the plant just before the seed is shed. A large number of plants shed their seeds explosively, the pod or fruitcase bursting to fling the seeds everywhere. These types are best collected a few days early and the cycle completed in a controlled way by inverting the seed heads in a paper bag suspended in a dry, warm place in the sun (Fig. 12).

Separating seed from chaff is fairly simple in most cases. When perfectly dry both seed and chaff should be placed in a shallow tray and gently rolled from side to side, at the same time blowing over them in the same gentle manner. Any unbroken seed cases can be rubbed between the fingers to extract the seed. Large pieces of

chaff and other waste are best removed with the fingers or tweezers as the process continues.

Seed contained within fleshy fruits such as rosehips should not be dried out, as germination can prove difficult when the seed is eventually sown. Such seeds are best sown as soon as they are collected, or stratified as described earlier.

In the spring the sand/peat mixture in which the seed has been placed should be turned over every few days with the fingers. When a few seeds are seen to be sprouting the whole mixture should be sown in an appropriate manner for the particular species.

Bulbous plants very often set seed freely, and those species flowering early in the year are amongst the easiest of all to collect. Unfortunately, it does require patience to see the results of your labours as they often take several years to reach flowering size. To offset this the enthusiast should collect and sow some seed every year, then after the initial few years of waiting subsequent seasons bring forth their results annually. Once more bear in mind that those plants bearing varietal names, such as 'King Alfred' daffodils, are hybrids and will not grow true from seeds.

F1 HYBRIDS AND HYBRIDISATION

The many hybrid plants in our gardens are generally the result of crossing one species of plant with another in the same genus, or between distinct varieties within a species. An example is the polyanthus, which originated as a cross between *Primula acaulis* and *Primula veris* (the primrose and the cowslip, the wild hybrid being known as the false oxlip).

Almost all modern strains and varieties of such plants are the result of selecting and/or crossing the seedlings which resulted from the original cross.

Hybridisation is a complex subject and the results from many crosses are unpredictable, which is why a great many seedlings have to be raised to find that one plant with all the virtues required. This, as well as the necessary trials and selections that follow, accounts for much of the cost of 'novelty' introductions.

The first seedlings to be raised from a cross between two plants are called the first generation, or F1. Subsequent generations of crosses from these F1 plants are labelled F2, F3 and so on.

Although F1 is a term used for all first generation seedlings it is particularly significant when attached to the F1 hybrids increasingly offered in seed catalogues. These hybrids are the result of crossing two parents with distinctive qualities, the first generation bringing out the best of both. The parents used for this type of seed

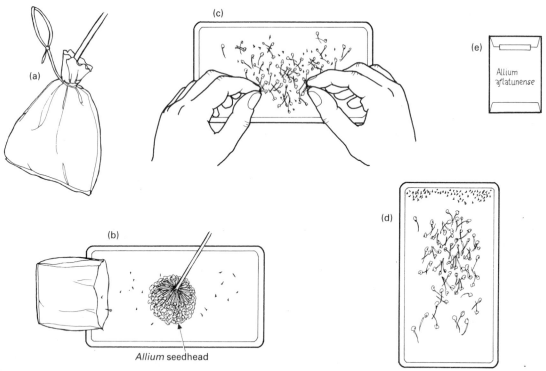

Allium seedhead

Fig. 12 Preparing seed for storage. (a) The seedhead tied inverted in a paper bag for final drying, and hung in a warm spot. (b) When dry, empty the contents into a small metal or plastic tray. (c) Roll the seedpods between the fingers to release the seed. (d) Shake the tray gently from side to side to separate the seed from the chaff. (e) Label clearly and store in a cool, dry place in unsealed paper envelopes.

production are plants which have been grown and selected over many generations until they are a pure line, this meaning that they always breed true from seed and produce virtually identical plants.

This process generally leads to a weakening of the plant characteristics, but, rather surprisingly perhaps, when two distinct pure lines are crossed they yield a first generation of extremely strong, often disease-resistant offspring which show very little variation between individuals.

Great skill is needed to raise F1 hybrids, as both parents often have to be isolated to prevent accidental cross-pollination, and in many cases pollination has to be done by hand under artificial conditions. This, and the care taken in selecting and maintaining the distinct pure lines which provide the parents, is once more reflected in the cost of such seed. Nevertheless, anyone growing plants from F1 hybrid seed can hardly question the value received in results.

Seed saved from F1 hybrids will provide a mixture of all the genetic forms within the plants, many of which will be completely useless. For this reason seed is not worth saving.

Occasionally, among such plants as chrysanthemums, plants appear either wholly or partly changed in colour from their varietal norm. Such plants are mutations or 'sports' which appear for no apparent reason. These will not breed true from seed but promising variations can be propagated vegetatively and many new varieties have been raised in this way. They are not to be confused with 'stray' plants of another variety.

SOWING SEED

Seed is sown in two basic media; either in a specially formulated compost for growing under glass or in the ordinary soil of the open garden.

In many respects sowing in compost under glass has more advantages, the main one being that sowing is less dependent upon weather conditions and plants started into life early in the year generally yield a better crop. With many plants it is essential that they are started into growth under glass, indeed, many are best grown in such conditions throughout their lives.

In all cases under glass the first essential for successful seed raising is cleanliness, as the main hazard facing seedlings in a warm, relatively still atmosphere is an attack by one or more of the various fungus diseases. The main offenders are those causing the condition known as 'damping off', symptoms of which are the rotting-off of the seedling at ground level and its subsequent collapse.

The risk is lessened by using sterile containers and compost, plastic being preferable for the former as it is easier to clean than other materials. Should soil be used in the compost then this too should be sterilised. A simple way of sterilising a small quantity is to fill a large saucepan with dust-dry soil, putting it in on top of 1 cm of water. Bring this to the boil, then simmer for fifteen minutes. Turn the soil out on a clean place to dry until it can be handled, before mixing into compost. For larger amounts various sizes and types of commercial sterilising units can be obtained or for the greenhouse and outdoor beds various chemicals can be used to good effect.

Strong healthy seedlings are less prone to fungus diseases, and such plants can be encouraged by sowing thinly and keeping the seed compost evenly moist without allowing it to become waterlogged. A preventive spray with either Cheshunt Compound or a liquid copper fungicide immediately prior to sowing, and again when pricking out, can be nothing but beneficial in the majority of cases.

It is important to use a good seed compost and there are many proprietary brands on the market or the more enterprising can easily make up his or her own. The John Innes formula for seed compost (see page 3) or soilless compost suggested for cuttings should prove adequate for the most part, but the plants must be pricked off without delay once they germinate.

Before use all compost should be evenly moist but not saturated unless culturally necessary.

Before making a start it is best to prepare properly by making sure that all materials are at hand and labels are written. Fill the pots or boxes with your chosen compost, filling them almost to the brim, and press down the corners and sides in the case of boxes to dispel any air pockets, settling the contents of pots with a couple of light taps on the workbench top. Finally, level the compost with the fingers and firm it lightly with a wooden 'patter' or the base of a clean pot (Fig. 13).

Sow the seed on this surface as evenly as possible, ideally by placing each seed individually with the fingers with the larger types or by sprinkling them from the hand between thumb and forefinger. Smaller seeds are best sown direct from the packet, the smallest being mixed with a little clean dry silver sand to make them more visible. It aids more accurate distribution if the top of the packet is cut across cleanly with scissors, and one side bent to a V-shape with the fingers. This forms a channel

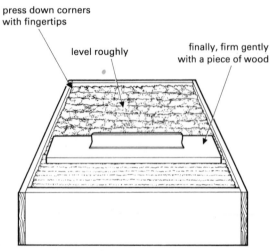

press down corners
with fingertips

level roughly

finally, firm gently
with a piece of wood

Fig. 13 Preparation of seedbox prior to sowing or pricking
out small seedlings.

through which the seed can be shaken in a controlled way.

Very small dust-like seeds should be merely firmed into the surface of the compost with a patter, whereas large seeds need to be covered in compost to various depths. A good general guide is to cover the seeds with the same depth of compost as their girth or thereabouts, but unless otherwise specified never sow seed too deeply or it may fail to germinate altogether (Fig. 14).

Some seed companies offer pre-sown seed—all you need to do is add water and wait. The more recent ones use a specially prepared vermiculite in a plastic container. If you find sowing a chore they are worth considering, but you pay more for the seeds of course, and the range available is small. You still need to provide sufficient warmth and light.

If it is the pricking out rather than the sowing that you find difficult, you can even buy seeds attached to a piece of card that you push into the compost at the right spacing (there is a depth mark on the 'stick' too). There may be several seeds on a 'stick' but you simply thin to one if more germinate. The drawbacks are the same as those for pre-sown seed, already described.

Compost used to cover the seed should be sifted over them carefully with the fingers taking care not to use too fine a compost as this is inclined to cake and prevent air reaching the seeds.

With the seeds sown, the pans and/or boxes are grouped together and given a gentle watering with a fine spray before being covered with sheets of glass, to conserve moisture, and in many cases with paper, to exclude very bright light. Newspaper is perfectly adequate for this.

In many cases no further watering will be required until after the seeds have germinated, and a daily watch is essential to note when this occurs. As soon as a few seedlings have appeared both glass and paper are removed taking care that the now unprotected seedlings are in good light but shaded from full sun.

Never leave seedlings covered with glass or paper once germination has begun or they will become drawn and leggy within a very short space of time. Such plants never make satisfactory mature specimens.

Once the seedlings are established they will need transplanting or 'pricking out'. The sooner this is done the better but I leave it until the first true leaves begin to appear when the seedlings can be handled with comparative ease.

Very small seedlings are best pricked out into the same type of compost as they were germinated in, feeding them as necessary as they become established. Stronger growing types need a stronger compost, J. I. Potting Compost No. 1 or its equivalent being ideal. The J. I. potting composts are based on the following formulae:

7 parts (by volume) medium loam (sterilised)
3 parts moss peat
2 parts horticultural grade coarse sand

For J. I. No. 1 add the following per bushel:

110 g of fertiliser made up of:
2 parts (by weight) hoof and horn meal
2 parts superphosphate of lime
1 part sulphate of potash
plus 20 g ground limestone

As with the composts, the fertiliser can be bought made up as John Innes Base Fertiliser. As a guide to compare with other suitable proprietary brands its approximate analysis is: 5% Nitrogen, 7% Phosphoric Acid and 10% Potash.

The same procedure as for preparing the seed containers is followed for pricking out, filling the pots or

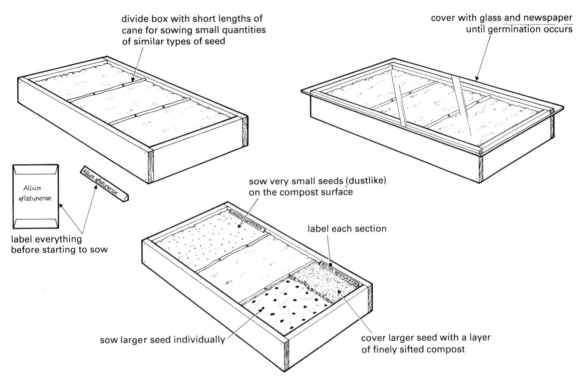

divide box with short lengths of
cane for sowing small quantities
of similar types of seed

cover with glass and newspaper
until germination occurs

Allium
aflatunense

Allium aflatunense

label everything
before starting to sow

sow very small seeds (dustlike)
on the compost surface

label each section

sow larger seed individually

cover larger seed with a layer
of finely sifted compost

Fig. 14 Sowing seed under glass.

boxes almost to the rim and firming them to exclude any air pockets. The compost used should be evenly moist and warmed to the greenhouse temperature, this being especially important in the early part of the year. It is easily done by preparing the compost the day before using it and leaving it on the greenhouse bench overnight.

Seedlings should be pricked out 2.5–7.5 cm apart depending on their strength of growth and size. Avoid disturbing the roots too much by teasing them out of the compost with a pointed stick. Always hold the seedling by the leaves and never the stem which is all too easily bruised. Small seedlings can be supported using a thin slice of wood with a V notch cut in its end, called a 'widger' (Fig. 15).

A hole is made to receive the roots with the same pointed stick as was used to lift them, taking care that the roots are not doubled back on themselves. Seedlings with exceptionally long roots can be made much easier to handle by dipping them in a small pot of mud made from sterilised soil and water. By giving them a twizzle as they are removed, the roots cling together whereupon they are easily twirled into the hole made to receive them. This technique is also useful outdoors for such plants as onions.

After pricking out, the seedlings should be kept warm, moist and lightly shaded until they are established, spraying them over daily with a fine spray if the weather is hot. If the original compost was sufficiently moist they should not need a real watering for a few days.

Seedlings which are destined to be planted outdoors to mature are grown on until the weather and season is right for them to go outside, hardening them off to outdoor conditions by gradually cooling them. This is best done by transferring them to a cold frame several weeks before planting out, gradually opening the frame's lights for longer periods until they are taken off altogether. Care is needed at this stage to prevent the plants being either baked in the hot sun or frosted by a sudden spring frost. The former can be prevented by shading, the latter by covering the cold frame lights with sacking or old carpet.

Plants to be grown in pots for either greenhouse culture or the home are potted on when they have outgrown their pricking out positions. Pots of 7.5 cm are usually used at first with John Innes Potting Compost No. 1 forming the growing medium.

As the season progresses such plants are repotted into stronger mixtures and larger pots, John Innes Potting Composts Nos. 2 and 3 being made to the same basic formula as No. 1 with only the fertiliser and lime content increased. With No. 2 this is doubled and with No. 3 trebled.

SOWING OUTDOORS

The main factor governing outdoor sowings is the weather. It is no use going ahead with sowing if the ground is cold and wet, far better to wait until there is a definite improvement and the soil warm and workable. There is nothing to be lost by waiting a while and much to be gained in better germination and stronger growth.

Seed can be sown in the open ground by either broadcasting it or sowing it in drills.

The broadcast method is usually reserved for lawns and the odd bare patches in the garden where it might be advantageous to sow a few annuals, perhaps between newly-planted shrubs which will take a few years to establish themselves.

As its title suggests the method is simply executed by sprinkling the seed in an even but haphazard manner on the soil surface, the process being completed with a light raking in opposite directions to settle the seed in a little. Birds can be a problem and are best kept at bay by

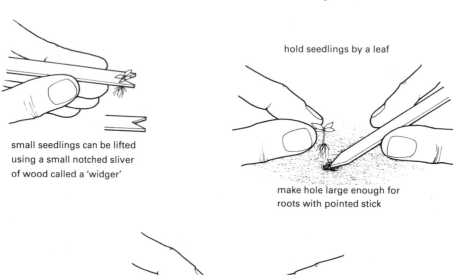

small seedlings can be lifted using a small notched sliver of wood called a 'widger'

hold seedlings by a leaf

make hole large enough for roots with pointed stick

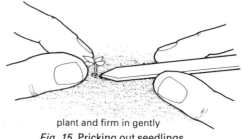

plant and firm in gently

Fig. 15 Pricking out seedlings.

stringing black cotton about the area. Some lawn seed is also treated with a bird repellent.

Sowing in drills is a more exact science, necessitating in the first instance a string line to make sure that the drills remain straight and parallel. If there is a place for neatness and regimentation in the garden then it is the vegetable and nursery plot.

Drills are made at variable distances apart and to various depths depending on the type of plant to be grown. It is usual to sow outdoor seed deeper than those under glass and the various distances and depths will be indicated in the plant lists.

Drills are easily made with the corner of a hoe or rake, drawing the tool slowly along by the side of the line, being careful not to move it out of position. A few extra canes placed along the line helps to avoid this on longer rows.

In some cases, such as pea rows, the drill becomes extended into a bed some 15 cm wide by 5 cm deep. This is best made with a spade, being careful to keep to the same depth throughout the length, as with the smaller drills.

Before starting any outdoor sowing programme the soil should be well prepared and broken down to a fine tilth with a rake or cultivator. If artificial fertilisers are being used then these should be worked into the soil a few days before sowing is to take place. If lime is also being added the two must never be put on together as they react unfavourably when in close contact. Incorporate one into the soil before adding the other.

This done and the drills made, the seed is sown thinly and carefully in the same way as sowing under glass. Again large seed such as peas and beans are best placed individually.

Once sown the seed is covered by gently drawing the soil back over them with a rake, and I always like to follow this by lightly firming the row with the back of the rake before moving on. With the line at its next station a row so marked is easily seen. Always leave a cane at each end of all rows as a marker, with a clearly written label. Memory is a most doubtful ally for the gardener.

Should the soil be dry then it is best to water the drills before sowing, preferably some time before to allow the moisture to spread. Never water the seeds in as this may cause panning of the soil surface which will effectively seal out the air necessary for good germination.

Once the seedlings are up and growing strongly they will need thinning. This is done by removing those plants which are not wanted, the final distance between the plants being determined by their type.

Thinning is best done in stages, the first when the plants are very small. This consists of taking out any obviously weak or unnatural specimens as well as reducing the numbers in those 'clumps' which always seem to appear in hand-sown rows. A week later a few more can be removed, again taking out the weaker plants but also bearing in mind the eventual spacing. Some time later a final thinning ideally leaves a row of evenly spaced individual plants.

Seed can also be sown in drills using a mechanical seed sower which is most helpful if large quantities are to be sown. They are generally fitted with an adjustable spacing guide which makes the line redundant once the first drill has been made.

SPORES

Spores are the reproductive units of the fungi, mosses, lichens and ferns which perform a similar function to that of seeds in the higher plants.

The spores of ferns are very small indeed, virtually microscopic, and are produced in vast quantities by many fern species. They are produced on the underside of the fern leaves or fronds, sometimes on special leaves adapted for spore production (Fig. 16). The clusters of sporangia are called 'sori'.

Most spores are easily collected by the amateur, the same principles as in seed-saving being applied. As spores are so light and easily dispersed when ripe it is important to keep a frequent watch on them as they ripen, this being indicated by a gradual browning of the sporangium which bears them.

As soon as this browning is pronounced the frond should be cut from the plant and placed in a clean paper bag for its final development, fastening the bag at the top to prevent the spores escaping.

Best results can be expected from freshly ripened spores and they can be sown at any time, although spring is probably the best. Greenhouse ferns will require a temperature of 18–21 °C for germination, whereas hardy ferns are happy with a little less.

As the germination of ferns is biologically complicated it can take some months for them to reach transplanting size. As the conditions favourable for fern spore germination are also favourable to other unwanted spores, everything used should be completely sterile. Once in sterile pots the compost should be treated by

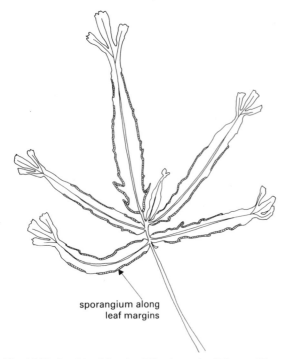

sporangium along
leaf margins

Fig. 16 Underside of fern leaf (*Pteris cretica* 'Wimsettii').

pouring boiling water over it, leaving it a few minutes to cool before sowing the spores.

A suitable compost is one consisting of equal parts by volume sterilised loam, moss peat, leaf-mould and sand; or the John Innes Seed Compost mixed with a further two-thirds (by bulk) moss peat.

Pans of 13 cm make suitable containers and should be filled almost to the top with compost, levelled and firmed and then treated with the boiling water. Sow the spores thinly and evenly over the surface of the compost (a job best done in good light and away from draughts) and press them down slightly into the surface with the base of a sterile pot. When finished the pot should be stood in a dish of water to maintain the compost in a moist condition, placed in a shady position, and covered with a sheet of glass, this being removed daily to wipe away the condensation to avoid drips.

After a number of weeks a green slime should appear on the compost surface. This is the first stage of the fern's growth. The slime is made up of large numbers of separate units called prothalli. These continue to expand for some time, and when their growth appears to have ceased pricking out can begin.

This is done by carefully removing 5 mm sections of the prothalli and transferring them to other pots prepared in the same manner as before. Space the pieces out on the surface of the compost about 2.5 cm apart and simply work them in, making sure that they are never planted any deeper than they were originally. The pots are then covered with glass and stood in water as before.

After about a week start to ventilate the pots by raising the glass slightly, giving them more air each day until by the time the new ferns can be seen they are ready to be left in the open air entirely.

When large enough to handle the young ferns are potted individually into small pots, potting on as soon as the pot is filled with roots, as most ferns resent being starved in their early stages.

SPAWN

Though mushrooms belong to the fungi group of plants and so reproduce naturally by spores, a vegetative method of propagation is used by both amateur and professional growers.

Mushrooms themselves form only a small part of the entire plant, being only the fruiting part to bear the spores. The main life of the plant is carried on below ground as a mass of white threads called the mycelium. It is with small sections of this mycelium that the gardener begins, this being supplied either in dry 'manure blocks' or inoculated into inert grains of rye or millet. Both types are known as spawn.

The production of spawn is very specialised indeed and is undertaken in sterile laboratory conditions. For this reason even professional growers obtain their spawn from specialist producers.

Traditionally, horse manure is the basis of mushroom compost, but nowadays there are also artificial methods of producing it, these usually from a straw base. Horse manure is treated by stacking and fermenting it for 10–14 days when the temperature within the heap should rise to around 65–71°C. When this is reached the heap is turned so that the outside becomes the centre, shaking it out and watering with a fine rose if it appears dry. Do not overwater.

The heap is turned twice more at 4 day intervals and at the second turning it is generally recommended that a certain amount of gypsum (calcium sulphate) should be added. The amount varies, but a rate of 1 kg gypsum per 50 kg of compost is reasonable.

After this time the compost is moved to the growing bed, treading it down in two layers to form a complete bed 15–20 cm deep.

The spawn is added when the compost temperature falls to around 21 °C, block types in plum-sized lumps 2.5 cm deep and 23 cm apart, grain types spread on the compost surface at a rate of around 305 g to 1 m^2 working it into the compost with the fingers and firming in.

About two weeks later the bed is 'cased' with sterilised soil or moss peat/ground limestone in equal quantities by weight to a depth of 3.8 cm.

Block spawn can also be used to supplement naturally-occurring mushroom colonies in lawns or fields by sowing small pieces just below the surface of the turf in warm, moist weather during summer; results, however, are unpredictable, depending as they do on the weather.

3 Division

The division of plants which are already established in the garden is probably the easiest of all methods of propagation and certainly one of the most successful for the beginner. As suggested by its name, the method involves the dividing up of an existing plant, each part divided being complete with its own roots and growth buds.

Division is most often done during autumn or early spring when the plants are dormant, and is used for the increase of many perennials, alpines, bulbs and houseplants. The great range of plants which can be divided naturally leads to a number of ways of achieving the desired result. It is easiest to divide these methods into six basic types, each dealing with a particular group of plants having the same growth characteristics.

FIBROUS-ROOTED PLANTS

Many perennials in the herbaceous border will be of this type; Michaelmas daisies, anthemis, artemisias and heleniums being just a few of them. After a few years in the border many plants will have overgrown their space and will possibly begin to die back in their centres. For their own good these need to be lifted and divided.

Though most plants can be divided in autumn or spring some are best done in spring only, pyrethrums and scabious being amongst them.

With small, loose-rooting kinds of plant it is a simple matter to separate the roots with the hands, pulling a small clump containing a few crown buds away from the parent plant (Fig. 17). Others may need the assistance of a sharp knife and some more drastic treatment still.

Should fingers or small knives seem inadequate when faced with a large and extremely tough root system, then a division will have to be made with the aid of two forks. Either hand forks or full-sized forks can be used, depending on the strength of the gardener and the resilience of the root system to be tackled. The method used is identical in all situations, the forks being driven through the centre of the clump back to back and then carefully levered apart.

Sometimes, especially with long established plants, it may be found that their centre has become woody and largely unproductive. In these cases the centres should be discarded, only the fresh new growth around the perimeter of the crown being replanted.

A mention might be made here about weeds. As many perennial weeds such as couch grass are the bane of the herbaceous border, it is often only at such

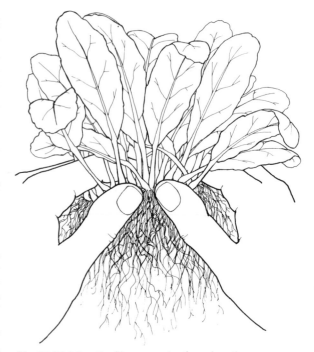

Fig. 17 Dividing the fibrous roots of a polyanthus.

moments as when plants are being divided that the troublesome individuals can be removed with any success and every effort should be made to do so before replanting the new divisions.

WOODY-ROOTED PLANTS

Some herbaceous perennials will be found to have very woody roots and crowns and may need rather more drastic treatment to effect a division than other types. Established plants of this kind should be lifted in spring before the plant has started into growth in order to carry out the division.

It helps greatly with the task if the crown of the plant is washed thoroughly once lifted. This enables one to see the buds easily so that the division can be made in the most favourable places. This can usually be done with a strong, sharp knife, the plant being cut up into appropriately sized pieces.

Should a knife or your muscles seem inadequate for the job then a spade or curved edging iron can be used.

Although treatment like this may seem somewhat harsh the plants will soon make good once planted in their new position. Less damage will occur if the cut is made without any hesitation. Just choose the best place to make the cut and give it one heavy chop. One enormous slicing cut is far better than several bruising half-hearted attempts trying to do the job gently.

OFFSETS

An offset is a young plant produced alongside the parent and easily detached from it. There are several kinds of offsets. Some cacti and succulents, for instance, produce a cluster of small plants around the base, and these offsets merely need separating and potting up. Bromeliads produce offsets around the old flowering shoot, and bulbs produce offsets freely.

Everyone who has had any dealings with bulbs will undoubtedly have seen smaller bulbs or bulblets growing from the base of the larger specimens. Daffodils in particular can be seen in any garden shop with several growing tips projecting from apparently one basic bulb. Such bulbs are called 'mother bulbs', and if the smaller bulbs growing from them are removed and grown on they will become individuals in their own right and eventually produce bulblets of their own. These are known as bulb offsets, a natural method of propagation which the gardener can make use of to good advantage (Fig. 18).

Some plants such as crocus and gladiolus produce corms which, although very similar to bulbs in appearance, are completely different internally. These are formed from a swollen stem, whereas bulbs are formed from modified leaf bases. With corm-bearing plants, new corms or cormlets develop on the top of the old corm or around its base and can be removed and grown on the following season as before.

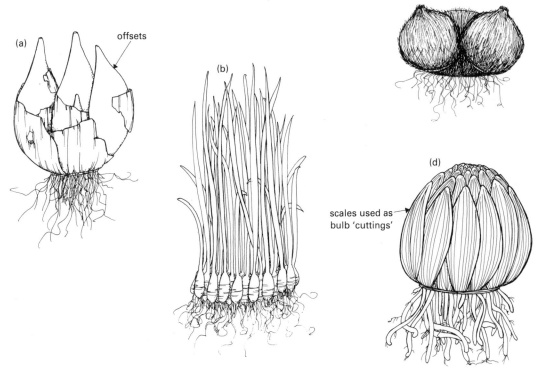

Fig. 18 Offsets forming on bulbs and corms. (a) Tulip. (b) Chives (*Allium schoenoprasum*). (c) *Ixia*: cormlets growing on top of the old corm. (d) *Lilium.*

Many bulbs and corms are lifted from the garden or pots during their dormant season and this is a convenient time to remove any offsets which might have appeared.

Liliums form a different kind of bulb again and can be increased by means of their scales which are broken off at transplanting time. It is debatable whether this method is really a form of division and should perhaps come under the heading of bulb 'cuttings', but nevertheless this is the method to use. John Innes Seed Compost is a suitable medium, the containers used not being filled to capacity. Cover the compost with a thin layer of silver sand, then stand the broken-off scales upright like cuttings, 2.5 cm or so apart, before filling over with sand until the tips of the scales just show above the surface. They can be dibbled into the sand with a stick or label if you prefer, but always leave some space in the box or pot for top dressing as the new bulblets form. The pots or boxes should be placed in slight warmth to encourage the formation of the bulblets. Do not overwater.

Many other bulbs can be propagated by means of their internal scales or layers and the enterprising amateur can gain much satisfaction from experimenting. Bulblets can be formed by mixing scales with a mixture of equal parts peat and silver sand, enclosing them in a polythene bag tied at the top, and incubating them in a warm place until bulblets have formed. These are then removed and grown on as for seedlings. Some, such as daffodils, may require a period of cold growing at around 4 °C following the incubation period to encourage rooting.

RHIZOMES

Quite a number of plants have underground root-like stems which, by storing food, enable the plant to survive during the winter and to propagate itself by vegetative means. The weed couch grass is one such plant, the underground stems spreading the plant far and wide, as many gardeners know to their cost. Happily, not all such plants are weeds.

Among the cultivated plants the bearded iris is probably the most recognisable of the rhizome-bearing types. The rhizomes of these plants are easily divided by cutting pieces off the underground stem of the parent plant after lifting. The pieces are formed by cutting sections 5–7.5 cm long from the parent plant, making sure that each part has at least one strong growing shoot rising from it (Fig. 19).

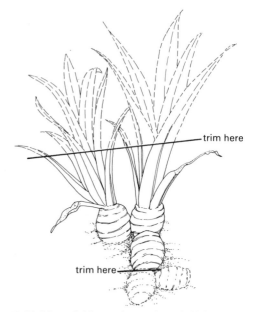

Fig. 19 Division of rhizome from a bearded iris.

With many rhizomes the best portions for propagation come from the outer edges of the plant, the original centre section being discarded. The retained portions are replanted in a horizontal position just below soil level. With most types the operation is carried out after the plants have flowered, usually around July.

SUCKERS

A sucker is the name given to the shoots sometimes arising as a type of offset from the root systems of certain plants and trees such as raspberries and poplar. These suckers often surface some distance from the parent plant and can be mistaken for seedlings.

Although more often than not a nuisance, these suckers can be used as a means of increasing stock, and are simply separated from the parent root and transplanted to a new position. This is generally done in autumn or early spring, the normal planting time for these types of plant (Fig. 20).

The job is usually simple enough, but lift the shoot carefully, cutting the sucker away from the original root with a piece of that root still attached to the new root base.

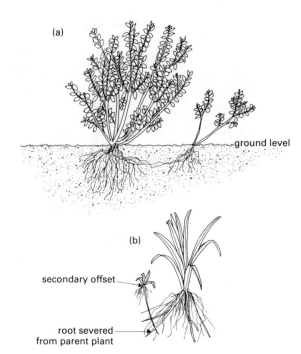

Fig. 20 Division (suckers). (a) A sucker growing from the roots of a forsythia. (b) A sucker/offset from a *Hemerocallis*.

Many plants can be increased in this way but it must be remembered that suckers rising from grafted plants such as lilacs and roses will produce plants of the *rootstock*, which is most unlikely to have the desirable characteristics of the grafted portion.

TUBERS

Tubers are formed from swollen underground stems or roots and are used by the plant for storing food, the commonest example of the type being the potato.

Tubers are quite simple to divide but as they come in a number of different forms handling them may, at first sight, seem more difficult than it really is. One essential in all cases is that, wherever the dividing cuts are made in the tuber, all pieces cut away should contain at least one crown bud or shoot.

Tuberous begonias can provide one example of division by this method, large vigorous tubers only being

selected for the treatment. Tubers are started into growth in the greenhouse in February as normal and grown on until the top growths are 1.3 cm or so long. The tubers are then carefully lifted and cut vertically through the middle with a sharp knife. A light dusting of powdered lime is then applied to the wound to discourage harmful organisms and prevent bleeding and the separated sections replanted. It is best after this treatment to place the plants in a warm propagating frame to establish them quickly.

Dahlia tubers, although looking much different from the tubers of the begonia, are again divided in a similar manner (Fig. 21). Once more choose only the largest and strongest for the treatment. As with the begonia, the dahlia tuber is cut vertically through the crown of the plant making sure that each section cut away retains at least one growing bud or shoot and a section or finger of the original tuber.

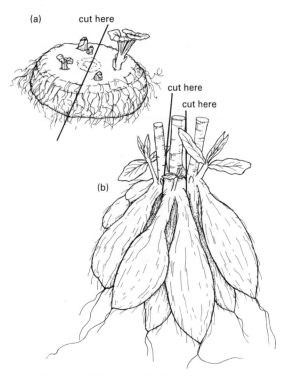

Fig. 21 Tubers. (a) Begonia tuber just starting into growth. (b) Dahlia tuber. Single stem tubers can also be cut by splitting through the centre of the stem.

4 Layering

Perhaps the best example of natural layering is the rooting of strawberry runners, or 'stolons'. All through the summer months the plants send out these runners from the axils of the leaves and they can cover quite a distance before the terminal bud produces a cluster of leaves and takes root. From the base of these new leaves arises another runner which moves on and so spreads or propagates the original plant rapidly and over a wide area (Fig. 22).

The gardener can take advantage of this natural method of propagation in the strawberry to increase his stock of plants. If a plant is required in an existing bed it is a simple matter to encourage a runner in the right direction and peg down the terminal bud where required. A few weeks later it will have sufficient roots to fend for itself and can be separated from its parent with a sharp knife, at the same time removing any runners forming from the new leaves.

Alternatively, a container of compost can be placed beneath a terminal bud and the bud pegged down to that. Eventually the new plant can be severed already potted up. As with all propagation material it is essential that the parent plant is healthy and free from disease.

To make an artificial layer similar to that of the strawberry is simplicity itself, nothing being needed in most cases other than a wooden or wire peg, a sharp knife and a suitable subject.

Very many shrubs such as magnolias and rhododendrons are likely candidates for layering, along with many plants in the rock garden. The method is also particularly useful for those plants not propagated easily by other methods, as the prospective new plant remains connected to its parent until rooting is well established.

During July and August a young, non-flowering branch should be selected. In the case of shrubs this will naturally be one close to ground level.

At a point where the branch will easily reach the ground, the sap is restricted either by giving the stem a twist, bending it or preferably making an incision with a knife just below a bud. The cut should be made towards the main stem 2.5–7.5 cm in length and just deep enough to form a tongue. Do not cut too deeply or the stem will probably break later. Before pegging down remove any leaves which may be beneath ground level when the layer is made.

In soil of good condition it might only be necessary to peg the branch down to the ground, keeping the tongue as open as possible without causing the branch to break, and then covering the cut with 2.5 cm or so of soil or compost. To assist in a satisfactory angle where the stem

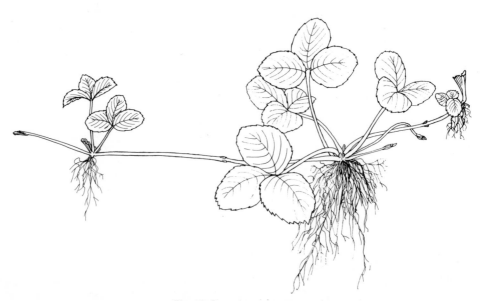

Fig. 22 Strawberry runners.

bends in the ground the leading edge of the branch can be tied in place to an appropriately placed cane.

If in doubt of the quality of the soil then a layer of some 5 cm of J. Innes Seed Compost or similar can be built up around the branch to be layered and the branch pegged into this, again covering with a further layer of compost over the top.

When sufficient roots have formed the new plant can be cut away from its parent. Root formation takes a variable amount of time depending upon the subject. Most shrubs take from six to twelve months, although rhododendrons and magnolias are better left for two years before separation.

Border carnations, which are also propagated by this method, should root and be ready for transplanting about eight weeks from being layered (Fig. 23). In all cases, should doubt exist as to the quantity of roots, leave the new plant attached to its parent for a further length of time. Shrubs, for instance, would benefit greatly if left to overwinter attached to their parent, separation and transplanting being left until spring.

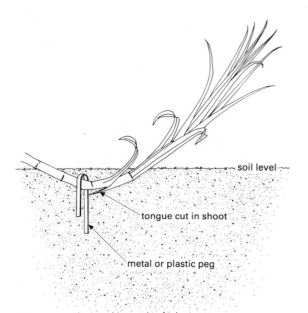

Fig. 23 Layering a border carnation.

soil level

tongue cut in shoot

metal or plastic peg

AIR LAYERING

Now we come to a more specialised form of layering, originally introduced from China and still widely known as Chinese layering. It is useful for propagating a wide range of shrubby plants and trees and enables roots to be formed on stems well above the ground without using soil at all (Fig. 24).

For all purposes the method of operation is the same. The materials required are a few handfuls of damp sphagnum moss (soaked in water and squeezed out), some waterproof self-adhesive tape and a tube or sheet of thin polythene.

Select a good, non-flowering shoot and make an upwards slicing cut into the stem where you wish the roots to form. The cut should be from 2.5 to 7.5 cm in length depending on the size of the shoot, and never deeper than half-way through. Should there be any danger of the stem breaking, then a short length of cane tied to the stem above and below the cut will act as a splint to prevent this happening.

Take a small amount of sphagnum moss and carefully insert it in the cut to keep it open, being careful not to break the stem. Dusting the cut prior to mossing with a hormone rooting compound may speed up rooting but is not essential.

With more moss build up a sausage shape around the cut, making the 'sausage' a good handful of moss. With outdoor work, especially, wrapping the moss around with cotton can save some frustration.

The final stage is to wrap the moss in polythene to exclude excess water when layering in the open air, and to retain moisture in all events. The thinner the polythene the better. With the polythene wrapped around the moss it is held in place and sealed with tape at the top and bottom in such a way as to prevent rainwater seeping in. Waterproof tape is, of course, essential.

If a polythene tube is used a slightly different procedure can be adopted, the tube being first slipped over the top of the stem and secured with tape below the cut. The tube is then packed tightly with moss, pressing it into position with a short stick or cane, and the top sealed as before.

After a period of time, dependent on the subject, the time of year etc, roots will be seen forming against the side of the polythene. When these appear well established the polythene can be removed and the stem severed below the sausage of moss.

Carefully remove all the loose moss but avoid

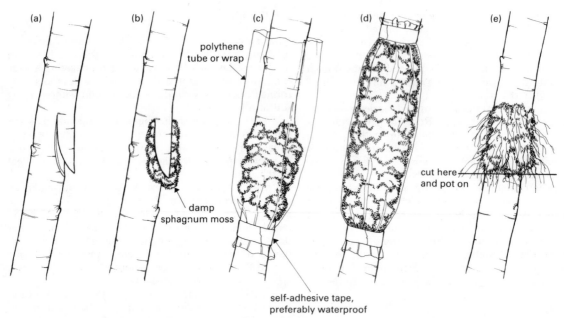

Fig. 24 Air layering. (a) Cut a tongue in the branch where roots are required to form. (b) Keep the tongue open with a little sphagnum moss. The cut in the stem can be treated with a rooting compound at this stage if desired. (c) Either surround the tongue with moss or slide a polythene bag over the branch and fasten in place at the base before packing tightly with moss. (d) Use a stick to push moss down into the polythene tube, and seal with tape at the top. (e) When roots appear on the side of the polythene, carefully remove the bag and any loose moss and separate the new plant by cutting off below the roots.

damaging the roots and simply plant or pot the newly-rooted plant into its correct environment.

Air layering is usually undertaken in spring or early summer when best results can be expected. As a general guide such plants as *Ficus* species (rubber plants) will form roots in some six to eight weeks under glass, whereas outdoor shrubs may take in excess of ten.

SERPENTINE LAYERING

This is a method particularly suited to long shoots such as those found on clematis and lonicera (honeysuckle).

Slanting cuts some 5 cm long are made with a sharp knife behind nodes at convenient distances along the stem chosen for propagation. These cuts are then pegged down into pots of sandy compost or directly into the ground, the intervening lengths of stem being left in the open. The pegged-down sections are covered with a

further layer of compost and left to root (Fig. 25). Clematis is normally propagated by this method in June, roots forming by autumn when the new plants are separated from their parents and either potted singly for protection through the winter or planted straight into their new growing positions.

TIP LAYERING

As with the layering of strawberries, this is a method exploiting a natural method of propagation, and it is very simple to undertake. It is used widely for propagating cultivated blackberries and loganberries, the tips of strong growing shoots being pegged down or dug into the ground where they touch. Roots will then form from the tip and by autumn the shoot can be severed and transplanted. Rambling roses, currants and gooseberries can also be propagated using this method.

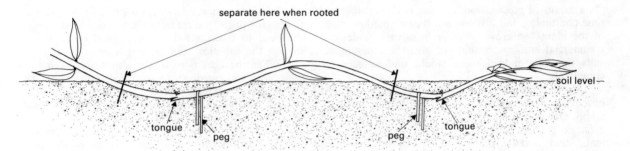

Fig. 25 Serpentine layering.

5 Grafting

As a means of propagation, grafting is rather slow and time-consuming, but if it was not for this method many of the plants we grow, whether in private gardens or commercial holdings, would not give the cropping results they do, indeed some would undoubtedly fail completely.

The principle of grafting is that of joining two parts of a living plant together so that they form a permanent union. One part of the plant (the rootstock) provides the eventual grafted plant with a strong root system and main stem, and the other part (the scion) forms the fruiting/flowering section of the plant. The scion is usually a section of stem taken from the previous year's growth of the parent plant.

By doing this, the fruiting qualities, for grafting is mainly used for the propagation of fruit trees, of choice 'cultivated' varieties can be combined with the rooting strength and vigour of the wild species or their selected strains.

As with the ever-increasing incidence of organ transplants in our human species, the two key words for successful grafting are compatibility and cleanliness. It therefore goes without saying that the rootstock and scion should belong to the same genus or plant group, although in a few cases plants within the same family can be successfully grafted together. It is also essential that both rootstock and parent plant should be disease-free and of vigorous growth.

There are many methods of making a graft and a great many more variations on those initial methods, but whatever graft is used the basic physiological principle remains the same: the cambium layer (*see* page 114) of the scion must come into direct contact with the cambium of the rootstock.

When the unions are made they are normally bound in position with raffia, or one of the more modern rubber, plastic or fibre bands specially manufactured for the purpose. In most cases, after binding, the union is sealed with grafting wax or a bitumen emulsion which serves the same purpose. Under glass, petroleum jelly serves reasonably well.

The grafting of fruit trees and ornamental shrubs is usually undertaken at the end of the dormant season in early spring. The scions are collected and prepared earlier than this, in December or January, tying them in bundles and storing them in cool soil against a north wall or fence. Commercially, special bins are made for this. In this way the growth of the scion is retarded until the sap is moving freely in the rootstock.

Scions are chosen from good stems of the previous year's growth and are cut from the parent tree at a length sufficient to contain four or five good buds. Before storage the soft tips are removed from each stem.

When the time is right, the scions are taken from their store and cut down to lengths of three buds. The preparation after this depends on which method of grafting is being used and these are outlined later under their various titles.

Fruit trees and ornamental shrubs are not the only subjects suitable for grafting methods and many woody perennials can be grafted along with such somewhat unlikely candidates as cacti and certain houseplants.

The methods of making a graft are so numerous and for so many purposes that I have dealt only with those most commonly in use and in my opinion closest to the propagating function. Few tools are needed for those grafts outlined here and most will be in the gardener's tool shed anyway. A strong sharp knife is essential for much of the work, with razorblades being suitable for the softwood candidates. Pruning saws and secateurs are useful in some cases.

SPLICE GRAFT

An ideal graft for the beginner, this is probably the simplest method of all. It is much used for grafting fruit stocks where scion and rootstock have a similar diameter. Its one limitation, however, is the need to hold them in position while the tie is made, there being no mechanical connection between the two parts.

Both stock and scion are prepared with a long slanting cut and the two cut surfaces brought together. Care should be taken to ensure that the cambial regions are in contact before finally tying and sealing.

Broom, roses and clematis can also be grafted using this method.

WHIP AND TONGUE GRAFT

This is a very similar method to the splice graft but with the refinement of a tongue of wood cut in both partners to hold them together while a tie is made.

The method is best suited to stock and scions of the same diameter, particularly those less than 2.5 cm. The scion is first cut with a slanting cut from its base extending some six times its diameter. A second cut is then made towards the top of the first in such a way as to form a downward pointing tongue. The stock is then

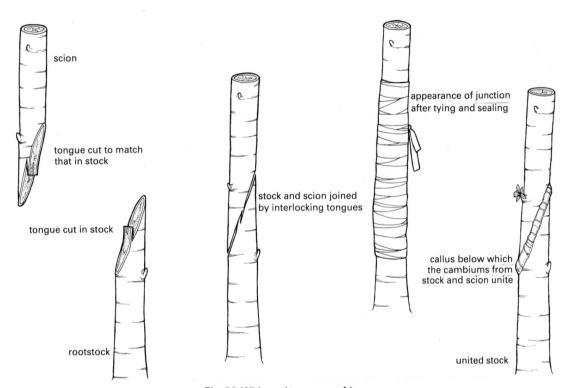

scion

tongue cut to match
that in stock

tongue cut in stock

rootstock

stock and scion joined
by interlocking tongues

appearance of junction
after tying and sealing

callus below which
the cambiums from
stock and scion unite

united stock

Fig. 26 Whip and tongue grafting.

prepared similarly with a corresponding slit made so that
when together the two tongues interlock. The graft is
finished with tying and sealing (Fig. 26).

This method is much used in fruit tree production in
spring using retarded scions as described earlier. As with
all grafts, any ties made are released by cutting once a
satisfactory union is seen to have been made.

WEDGE OR CLEFT GRAFT

The wedge graft has widespread uses and in its simplest
form is easily made. The rootstock is prepared by cutting
it off at an appropriate place and making a downward cut
into which the scion is placed. The scion in its turn is cut
to a thin wedge at its base and slid into the downward cut
in the rootstock making sure there is cambial contact at
one side at least. The joint is lastly tied with raffia, tying
the scion down with a loop over the first leaf if it shows
signs of being forced out by the tying process (Fig. 27).

The method is also used in its most drastic form in
topworking trees when the appropriate wedge slits are
made in the stock with a grafting tool or axe. The
variation known as oblique wedge grafting is to be
preferred in this case, the slits in the stock being made in
such a way that they do not extend fully through the
branch as would be the case with the true cleft graft.

Where herbaceous or in-leaf scions are used the
grafted plants should be confined to a warm close
atmosphere until a union has been made.

SADDLE GRAFT

This method is often used for grafting plants which are
difficult to propagate in any other way, such as certain
forms of rhododendron which are grafted onto rootstocks of *Rhododendron ponticum* (Fig. 28).

For a successful union it is important that the stem
of the scion and rootstock should be much the same size,

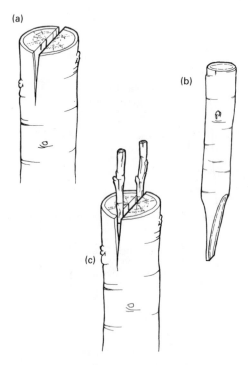

Fig. 27 Wedge or cleft graft. (a) A cleft cut across the stock. (b) The scion cut to a wedge shape. (c) Two scions placed opposite in the cleft.

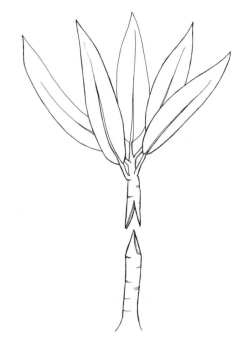

Fig. 28 Saddle grafting.

the rootstock for the purpose usually being a two or three year old seedling grown on in a pot for its last year in the case of rhododendrons.

The base of the scion is prepared by making two cuts slanting inwards to produce an inverted V, the knife being turned sharply at the end of each cut to form a kind of saddle. The stock is then worked in a similar way to take the saddle, the two parts being matched as closely as possible. Again, in the case of rhododendrons, the graft is made as close to the roots as possible so as to be underground when eventually planted out. In this way the scion can form some roots of its own, desirable in this case.

The graft is generally done under glass and as with other subjects done in this environment the grafted plants are best placed in a warm close atmosphere until a union has been made.

APPROACH GRAFTING

This method is in effect only an artificially induced 'natural' graft, the same as can be found on ivies growing wild where two stems crossing one another have grown together during the years. The method does differ from other forms of grafting in that the stock and scion are joined and established before the scion is severed from its own roots.

In its basic form the stock and scion are generally grown in separate pots and the two brought close together when the graft is to be made. With the two plants as close as possible, an appropriate place on the stem of each plant is chosen and a shallow sliver of bark and wood is removed from each, thereby exposing the cambium layer beneath (Fig. 29).

The sliver should, as far as possible, be much the same size from each stem so that there will be a good join between the two. As both plants are on their own roots, the graft can be done at almost any time, but is best when the sap is running freely in spring or summer.

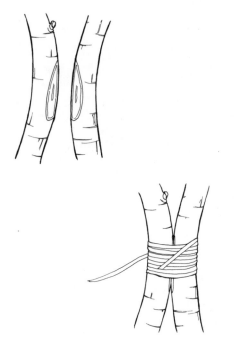

Fig. 29 Approach grafting. Detail of knife cuts and union partly tied in position.

With the two cuts made, the stock and scion are brought together and tied in place (Fig. 29), making sure that the cuts match as closely as possible. When the union has been formed satisfactorily, the top of the rootstock is cut away above the graft and the root of the scion below, this being done gradually over a number of weeks to lessen the shock to the plant.

Vines can be grafted by this method and it is also useful for subjects difficult to graft as both sections are able to sustain themselves on their own roots until a union is made. This type of graft is usually done under glass.

FRAMEWORK GRAFTING

Framework grafting is a general term which covers a number of grafting methods used to rework or convert the head of established trees from one variety to another. It differs from 'topworking' in that the main framework of the tree is retained, only the lateral shoots and spurs being replaced. It is the better of the two methods, frameworked trees coming into fruit production in less time than those which have been topworked.

To prepare a tree for frameworking, the side branches should be thinned out if necessary to shape the tree nicely and then all the smaller laterals and spurs are pruned away completely from those remaining, except where the tree is to be worked by stub grafting. These branches are then replaced by scions of the required variety, making the grafts every 20–25 cm apart (Fig. 30).

The scions used are longer than normal, bearing a minimum of six buds or as many as seven or eight. As the object of frameworking is to make the tree appear as natural as possible great care should be taken in placing the scions, remembering that scions grafted vertically on the inner or top surface of a branch will grow stronger than those in a horizontal position below.

Various methods of grafting are used in frameworking, stub grafting being one (Fig. 31). With this, the tree is prepared as before except that lateral growths of between 6 and 25 mm are left intact wherever a new growth is required.

The scions are prepared with a wedge-shaped base, one side being slightly longer than the other. A cut is then made in the branch to be grafted some 1 cm away from the main stem, making the cut towards the base of the branch but never deeper than its centre.

The branch is bent slightly to open the cut and the wedge of the scion placed in it, longer side of the wedge down. Once released the branch will hold the scion tightly, making it unnecessary to tie it in place though a few turns of tape around the graft would do no harm. With the graft made satisfactorily the original branch is cut away immediately beyond the graft and the job completed with a sealing compound. When the tree is finished the terminal branch on each main branch should be removed immediately above the topmost scion.

If stub grafting alone is insufficient to give the tree a balanced framework the gaps can be filled using a method called side grafting. This is accomplished by preparing the scion with a wedge-shaped base, one angle of the wedge being more acute than the other. This wedge is then inserted into a cut made in the side of the branch, the cut going a quarter of the way through. Both wedge and cut are made in such a way that, when finished, the scion will protrude from the branch at as natural an angle as possible.

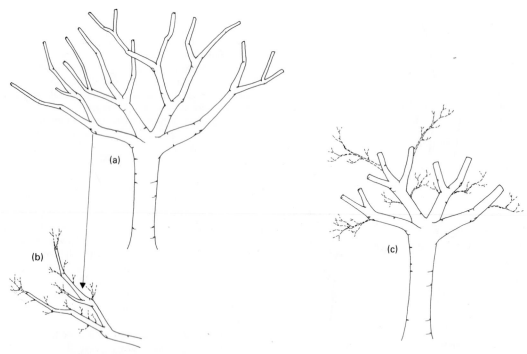

Fig. 30 (a) An apple tree prepared for frameworking. (b) Detail of a worked branch with the scions shown in dotted lines. (c) The same tree prepared for topworking. The sap-drawing branches are indicated by dotted lines.

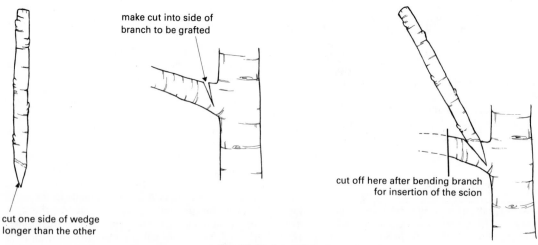

make cut into side of
branch to be grafted

cut one side of wedge
longer than the other

cut off here after bending branch
for insertion of the scion

Fig. 31 Stub grafting.

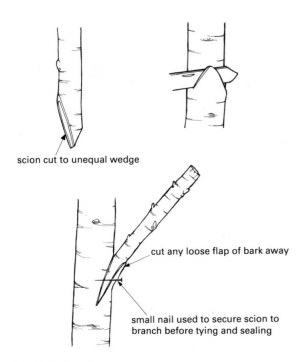

Fig. 32 Bark grafting.

scion cut to unequal wedge

cut any loose flap of bark away

small nail used to secure scion to branch before tying and sealing

Once again the cut is opened for insertion by bending the branch slightly so that when released it will grip the scion tightly.

Other methods used in frameworking include bark grafting (Fig. 32) and oblique side grafting, and in all cases methods can be mixed if doing so will yield a better over-all result.

It should be remembered that more than one variety can be worked on one tree, which is useful when cross-pollinators are required. It is useful, too, for those species having male and female forms, such as *Ilex aquifolium*, the common holly.

CROWN OR RIND GRAFTING

This method is used mainly for the topworking of established trees when it is desired to change one variety to another. It is best carried out in late April in most parts of the country, the scions being prepared earlier and stored and the tree to be worked being prepared by cutting away most of the main branch system some

60–90 cm above the crutch (the part of the tree where the branches separate from the trunk). A few smaller branches are left just below the main level of grafting to act as sap-drawers. These feed the roots and sustain the tree until the new branch system is established (Fig. 33).

The larger the tree the more sap-drawers are retained, these being removed or worked in their turn when the new wood is growing well.

Where possible, branches should be removed above a fork, making for more grafting, but lessening the chance of infection. For this same reason, the tree is best prepared and grafted in one operation.

Before commencing the graft, the cut on the branches to receive the scions is pared smooth with a knife, and the scions are then inserted under the bark or rind at the top of the branch. To do this, the scion is prepared with a flat slanting cut at its base, extending

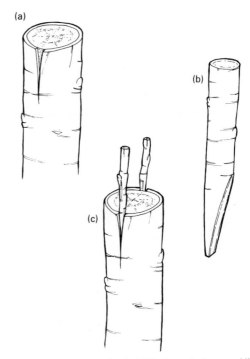

Fig. 33 Crown or rind graft. (a) Bark or rind cut and lifted from the wood. (b) A scion cut to fit is pushed between bark and wood. (c) Two scions placed opposite.

some six times longer than the diameter of the scion. A corresponding cut is then made down the side of the receiving branch from the top and the scion is pushed down between the rind and the wood, the cut surface of the scion facing inwards (Fig. 34).

A varying number of scions are inserted into a branch depending on its diameter, two to four being the general rule. Once made, the graft is bound and the whole wound encased in grafting wax or similar.

Many other methods are used for topworking other than the crown graft, cleft and veneer grafting being among them.

Topworked and frameworked trees should not be neglected after grafting and must be treated in much the same way as fruiting trees to keep them free from pests and diseases. Care in the early stages will pay dividends. In situations exposed to strong sun it may also be advisable to whitewash the trunks of topworked trees to prevent damage by strong sunlight, a condition known as sunscald.

Fig. 34 Position of hands when cutting a scion to fit the cut in the stock.

As might be expected with such drastic attention the trees will throw out many suckers and most should be removed, especially when they appear around the scions. A few can be retained and will act as sap-drawers for the first season or two. As with all fruit growing, attention to good pruning in the following years is necessary for good results.

DOUBLE-WORKING

Although not likely to be used by many amateur gardeners, this is still an interesting method of grafting to consider as its application is quite widespread.

Double-working is used for various reasons. The main ones are to overcome incompatibility between stocks which in theory should be compatible, and to build trees with strong straight stems, a process known aptly as 'stem-building'.

The first situation is commonly found with pears which are grafted onto quince rootstocks. Some varieties take well to the graft, but others, though perhaps developing an effective union for growth purposes, are weak in structural strength and subsequently break off as the tree becomes established. To overcome this an intermediate scion known to be compatible with both partners is grafted between the two, hence the term 'double-working'.

Even though this intermediate stem may only be a few centimetres long it can have a profound effect on the finished tree, and this effect has been successfully used to produce disease resistance and frost hardiness in top scions not known for these qualities.

Double-working can be used to produce straight stems when the variety you wish to grow is inclined to produce a bush rather than the desired tree. Using a mutually compatible scion of a good stem-producing strain, the intermediary is grafted to the rootstock first and the top scion added at that part of the stem where the head of the tree is desired to form.

Double-working is usually carried out in one operation, the scions and stock being prepared in the normal way. The intermediate scion is cut about 13 cm long and is grafted to the usual three-bud top scion. Whip and tongue grafting is perhaps best for this. After tying and sealing, the intermediate and top scion can be grafted onto the prepared rootstock using the same method.

Shield budding (*see* Chapter 6) can also be double-worked when budding pears, a small budless shield of an intermediate being inserted between the rootstock and the true bud shield as the operation is done.

ROOTSTOCKS

A description of all the rootstocks and their various methods of propagation and their specific uses would take a chapter of its own, so this section contains merely an introduction and one method of propagation peculiar to rootstock production. More information will be found under the individual plants in the plant lists.

Rootstocks can be raised in many ways: by seed, cuttings, layering and stooling.

Most methods are outlined elsewhere, but stooling is dealt with here. With this, a parent rootstock is planted and grown on until it is well established, then during the winter it is cut off near the ground. Treated in this way it responds by sending out a number of shoots from its base. When these are 12–15 cm high they are earthed up, continuing throughout the year as the shoots grow until the shoots are buried in 15–20 cm of soil. This keeps all light from their bases and they form roots during the summer.

In late autumn the soil is removed and the rooted shoots are broken away from the original rootstock now called a stool. Plants treated this way will yield a crop of shoots annually.

Much work on rootstocks has been done at the East Malling Research Station in Kent and the strains developed there and jointly with the John Innes Institute at Merton (before it moved in 1949) are used generally for the production of fruit trees. Taking apple stocks as an example, these are identified with numbers corresponding to their growth characteristics such as M (Malling) 9 and MM (Malling/Merton) 106, both rootstocks producing dwarf trees. Other rootstocks produce medium-sized and vigorous trees.

Prior to working, most hardy fruit stocks are planted out in the spring of one year and are grafted in the March or April of the next. It does no harm at all to leave them an extra year if they do not appear to be strong enough. Plant out leaving 38 cm between the plants in rows 90–120 cm apart if your planting extends to such numbers.

The whip and tongue graft is most used in this situation, the rootstocks being cut down to within 7.5–10 cm of the ground at the time the graft is made.

6 Budding

The method of propagation known as budding is really a form of shield grafting. It has, however, developed into a branch of propagation in its own right and is extensively used for the propagation of roses, although many fruit trees and ornamentals are also increased by the method. The technique used is much the same in all cases, the only real difference being the height at which the bud is placed on the stem of the rootstock, so I have chosen to describe in detail only one example, that of budding roses. More details of rootstocks etc of other plant types will be found under their respective names in the plant chapters later in the book.

Although some roses, particularly those nearest the original species (i.e. *Rosa moschata*, the Musk Rose), can be raised from cuttings, many modern varieties or cultivars are not strong enough to perform on their own roots to any satisfaction. To overcome this problem such varieties are budded onto various strong growing rootstocks which, generally speaking, are wild species of rose or selected strains of wild species. The object of the graft is to transfer a well-developed bud from the cultivated variety to the rootstock.

The main method used in the budding operation is quite simple to accomplish, although as with many things only practice makes perfect.

The first essential is to undertake the graft when the plants are in full growth. The sap must be travelling freely within the stem to allow the bark to rise from the wood of the rootstock without damaging the vital cambium layer, this being the narrow layer of growing tissue between the bark and wood of most plants. It is this cambium layer which is responsible for healing wounds, forming callus on cuttings prior to rooting, and for the uniting of budded or grafted stocks. In normal seasons plants are in full growth between mid-June and the end of August. Obviously those gardeners in the more southern areas will have a longer choice of season than those in the north.

The only tool needed for successful budding is an extremely sharp knife. Special budding knives can be bought and are best suited to the job. These knives have a single blade no longer than 5 cm in length and a 'bone' handle, the handle being tapered to a flat point to assist in the raising of the stock's bark.

As I have said, a knife designed for the purpose is naturally the best tool for the job, but as long as the operation is carried out with care any knife of a similar design can be used. A suitably shaped piece of wood or plastic can be used to lift the bark, or handymen may even be able to modify or manufacture a budding knife of their own. A typical design is shown in Fig. 35.

The shoot selected to supply the bud should be a good one which is carrying a flower near to shedding its petals. The shoot should be cut some 15–20 cm in length so that it carries several buds or 'eyes'. These eyes are identified as small knobs in the junction where the leaf stalk joins the stem. The eyes should be plump and firm but not yet started into growth, and the best eye will probably come from the centre of the cut stem.

The thorns on the shoot are also good indicators as to the hardness of the stem and can be used in determining the readiness of the buds for budding. The idea is to try to break off the larger thorns by pressing them sideways with the fingers (Fig. 36). If the thorn tears off the stem with shreds of bark attached it is not ready and should be left another week or so. If, on the other hand, they snap off cleanly, the stem is ready to use. The scar left by the removal of the thorn should be of tender green tissue. An over-ripe scar will appear hard-looking and dry. A stem of this type is also unsuitable.

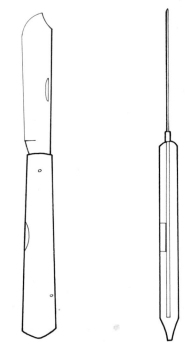

Fig. 35 A typical budding knife.

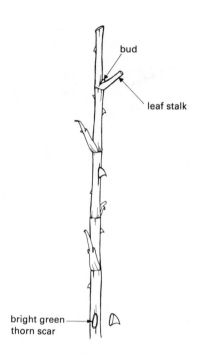

bud

leaf stalk

bright green
thorn scar

Fig. 36 A prepared bud stick.

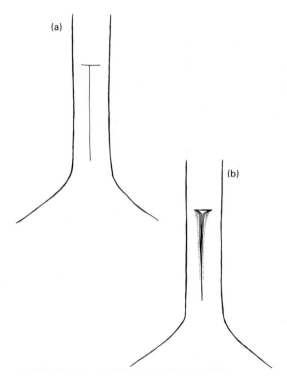

(a)

(b)

Fig. 37 (a) Position of the T-cut on the rootstock. (b) Bark raised to form two flaps.

Once a stem is selected it can be removed from the parent plant at the same time cutting the leaves off, leaving 2.5 cm of leaf stalk attached to the stem to protect the bud and allow for easier handling. At all stages drying out of the stem is to be avoided and prepared stems should be placed in a glass or jamjar containing a little water to prevent this.

The next stage is to prepare the rootstock. This is done by removing the soil from around its stem as close to the roots as is practicable and wiping the resultant exposed area with a rag to clean it.

A T-shaped cut is then made just below the original ground level (Fig. 37). This cut should be no deeper than the bark, with the downstroke of the T-cut being about 2.5 cm in length. Once the cut is made the bark is raised carefully from the stem to form two flaps. If the rootstock is in prime condition the bark will lift cleanly. On no account should it be forced as this would damage the cambium layer and possibly cause failure of any graft made.

With the T-cut made, press the flaps back against the stem to avoid any drying while the bud itself is prepared.

This is done by cutting the bud or eye from the prepared stem complete with a shield of bark. Start the cut about 2.5 cm below the chosen bud and gradually cut deeper until the bud is reached, whereupon the cut is made shallow again to allow the knife to emerge around 2.5 cm above the bud and, naturally, on the same side of the stem. What is then produced is a shield of bark containing a bud, a leaf stalk, and at the back a thin sliver of wood (Fig. 38).

The next step is perhaps the most difficult as the sliver of wood has to be removed from the shield without causing any damage to the root of the bud behind it. The beginner would perhaps be best advised to practise on a few rejected buds which can be thrown away with no loss before the real buds are attempted.

There are a number of methods of removing the wood from the shield, but the one I prefer is used as part of the entire operation. It is effected by making the cut as

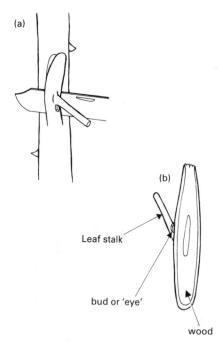

(a)

(b)

Leaf stalk

bud or 'eye'

wood

Fig. 38 (a) Cutting the bud shield from the bud stick.
(b) Detail of bud shield.

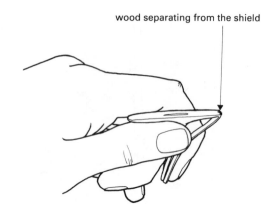

wood separating from the shield

Fig. 39 Removal of wood from the bud shield.

before, except that the knife is withdrawn before the blade reaches the end of cut and the job is completed by tearing. This results in the shield terminating in a long strip of bark. The wood is then removed in two stages. Firstly, it is loosened from the long tail of the shield by waggling the latter to and fro for a moment. Next, the shield is held by the leaf-stalk between the thumb and index finger of the left hand, the cut surface uppermost and the tail of the shield lying along and in line with the thumb (Fig. 39), the second finger of the hand holding it in place. This holding will cause the wood to rise slightly from the tail so that it can be grasped by the fingers of the other hand. By flicking the wood upwards and backwards and increasing the bending motion as the bud is reached, the wood is removed.

With the wood gone, the bud root should appear as a definite plumpness behind the bud on the shield. If, however, the area appears hollow it will mean that the root of the bud has been torn out with the wood making the shield useless. A few experiments will enable the

beginner to judge what is correct. Take heart from the fact that the written instructions are far more complicated than the actual handling of the plant material!

After giving these instructions it has to be said that not all growers advocate the removal of the wood from the shield, saying that its retention only rarely results in a failure of the graft. This may very well be so, but if you should decide to leave the wood on the shield it is essential to make your cut below the bud as shallow as possible so reducing the thickness of the wood behind the eye.

With the bud shield and rootstock now prepared, the base of the shield is trimmed with a cross-cut 13 mm below the bud. Once more lifting the bark at the T-cut in the rootstock, the shield is carefully pushed down the cut as far as it will go. For this purpose the leaf-stalk makes a convenient handle and also ensures that the eye is facing outwards in its correct position. With another cross-cut the top of the shield is trimmed to fit neatly in place at the top of the cut. The bark is then pressed into place around it. Finally, the wound is bound firmly yet not tightly with moist raffia, leaving only the bud and leaf-stalk exposed (Fig. 40).

After three weeks or so it will be seen whether or not the bud has taken. Both the bud and the shield should look bright green. If they look brown or black they have failed and another attempt will have to be made on the opposite side of the rootstock and preferably still lower down the stem.

If the bud has taken the tie should be removed and the plant left to grow on as it is until the following

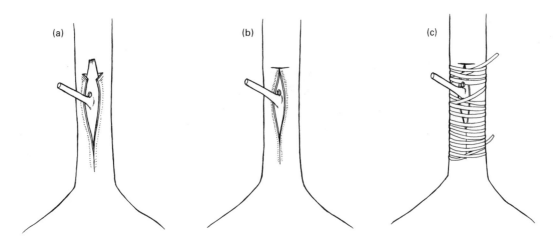

Fig. 40 (a) Bud and shield inserted in the rootstock. (b) Bud and shield in finished position. (c) Bud and shield fixed in position with moist raffia.

February. At this time the stock plant should be 'headed'. This means that all the top growth of the original rootstock is removed from about 13–25 mm above the budded eye.

The new growth will sprout out later in the spring and it may be necessary to tie it to a small cane on the opposite side of the stock to encourage upright growth. New plants will flower some time later than established plants and after flowering should be cut back fairly low to encourage new growths from the budding point. The new plant can be transplanted to its permanent position in the autumn.

CHIP BUDDING

A brief mention may be made here of another method of budding which is said to be easier and just as successful as the traditional method. The operation involves the removal of a chip of stem from the rootstock and the replacement of this by a similar piece from the mother plant containing a bud, as with ordinary budding.

This is effected by making two cuts across the rootstock 25–40 mm apart and removing a chip of stem from between the cuts with a knife. This is replaced with a chip from the mother plant, the bud being near the centre of the replacement chip. When in place it should be tied securely with raffia as before.

ROOTSTOCKS

There are many different rootstocks which can be used for budding roses, the type chosen depending on the soil the plants will be grown in, the rose to be budded and the rootstock most readily available to the prospective budder.

These various rootstocks can be raised in two ways; either by cuttings or from seed. They can also be bought as maiden plants (one year old), from certain nurseries who specialise in rose-growing. Generally speaking, seedlings are to be preferred as they provide the better root system. For both methods of propagation, *see* Chapters 1 and 2.

The most readily available rootstock for the amateur is probably the wild dog rose, *Rosa canina*. Stock from this species, either from cuttings or seed, is best suited to growing on heavy soils. Commercially, many strains of *R. canina* have been selected, but if the species grows wild in your area it is a good indication that stock from such plants will do well in your garden also.

For light sandy soils, except those over chalk, it would be best to use *Rosa rugosa* or *Rosa multiflora japonica* (syn. *R. polyantha multiflora* or *R. polyantha simplex*), which also does well on shallow sandy soils. For light soils of a calcareous nature *Rosa laxa* is to be recommended. This species has a thin bark and is easy to bud; its main fault is its tendency to throw up suckers.

(These can, however, be used to provide more root-stocks.)

Bush roses are budded onto two-year-old plants, mainly of *R. canina* or its numerous strains. Standard roses need to have a rootstock with a stem some 120–150 cm in height. It is possible to find such stems growing as suckers from hedgerow plants in the wild, but commercially standards are budded onto stems of *R. rugosa* or *R. polyantha simplex*.

With standards, the method of budding is the same as for bush roses, except that it is carried out at the top of the stem rather than near the roots. Usually, three buds are inserted on either side of the main stem at some convenient height, in the case of *R. rugosa*, or on the side branches growing from the three buds left to grow on the original cutting in the case of *R. canina* (Fig. 41).

When planting cuttings for rootstocks of either bush rose or standard production it is essential to remove all buds except the top three prior to insertion of the cutting. This will hopefully avoid a mass of suckers and unwanted sidegrowths on the eventual budded plant.

Rootstocks of all types intended for budding should be planted out in their budding positions in the autumn preceding the budding period. Needless to say, the area chosen for planting out should be in good light and the soil cultivated to a satisfactory growing condition.

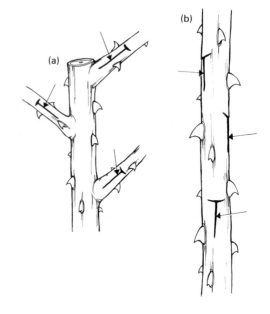

Fig. 41 Budding standard roses, showing the position of the T-cuts. (a) *Rosa canina*; (b) *Rosa rugosa*.

7 Greenhouses, Frames and Propagators

Without doubt, the owner of a heated greenhouse with a propagator and mist unit is by far the best placed to propagate the maximum number of garden subjects. Provided that he has the space, there is little he could not consider trying with a reasonable chance of success. Unfortunately, not everyone has these facilities, but they are by no means essential to an interesting propagating programme.

To begin, I will take as an example the lot of the average gardener whose interests and financial abilities evolve over the years. So we start with a house and a garden and little else except perhaps enthusiasm.

At once we can see that the range of plants which can be propagated without any protection at all is quite large. The seeds of most vegetables, hardy perennials, biennials and annuals are sown directly into the open ground during the warmer parts of the year. To these may be added various hardwood and softwood cuttings which will certainly root in a sheltered part of the garden, and among other methods of propagation layering, division, budding and grafting all have their subjects suitable for working in the open air, indeed the majority of such methods are carried out there.

Yet the average gardener still dreams of better things, so he buys a few cloches. These are best described as portable low frames and nowadays come in a variety of shapes and sizes from the bell jar to the polythene tunnel cloche. Many seeds can be sown much earlier if given cloche protection and either grown there to maturity or uncovered later as the weather warms. Cloches can also be used to protect hardwood and semi-hardwood cuttings as well as plants raised in the greenhouse and in need of hardening off to an outdoor climate (Fig. 42).

One step up from the cloche, though in no way superseding it as all additions to plant protection are complementary, is the cold frame. These are to be had in various proprietary designs, but the handyman can easily construct one of his own. The walls can be made of wood, brick or concrete, the top being covered with a frame light or lights of glass, polythene or clear plastic. The size is merely a matter of convenience to the individual, but a frame 1.5 × 1 m is a useful size. The walls are best made about 45 cm high at the back and 30 cm at the front.

As with greenhouses, frames should be sited in good light away from the direct shade of trees or buildings, though some measure of protection from cold winds is to be desired. Greenhouses are generally recommended to be erected to run from north to south, but it may be

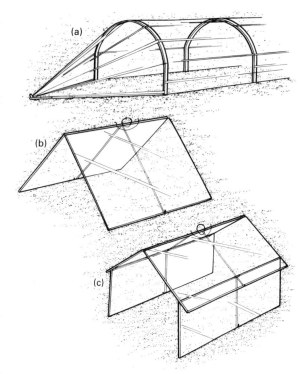

Fig. 42 (a) Polythene tunnel cloche. (b) Tent cloche. (c) Barn cloche.

noted that a greenhouse or frame running east/west will receive more light in the winter period and so is best suited to propagation purposes.

The cold frame is a very useful addition to the garden either for raising plants in its own right or as a transit post for hardening off plants raised in a greenhouse.

The usefulness of a frame can be greatly increased by the addition of underground mains voltage heating cables which are available in various wattage ratings. Naturally an electricity supply to the frame is essential and should be fitted using outdoor grade cable and fittings, preferably by a professional electrician, but in all cases to a high standard of safety and workmanship.

The heating cable should be buried under some 5 cm of sand, being careful not to let it cross over itself or touch. As a rough estimate a frame will require a cable size calculated at 6 watts per 100 cm^2 of floor area plus a further circle of cable placed around the side of the frame just above the sand surface calculated at 15 watts per 100 m^2 of glass or polythene above. For best results the cable is generally thermostatically controlled.

From the heated frame it is a simple but expensive step for a gardener to buy a heated greenhouse. This can of course be adapted for propagation only, when a greenhouse with half walls is to be preferred giving a better degree of insulation. Most gardeners, however, will want a greenhouse to serve a variety of purposes, and a propagating frame built within the greenhouse can serve them well.

Heated cables buried in sand are best for the indoor propagator too and an area similar to the frame outside can be constructed, the materials needed naturally being much lighter. Careful construction will allow it to be removed or adapted when not in use for its purpose (Fig. 43).

An alternative is the professionally constructed propagating frame widely available made in plastic. Various types are available from the simple covered box, consisting of a seedbox-sized base with a plastic dome, to the full thermostatically-controlled heated frame with variable ventilators. The prices quite naturally rise with the degree of sophistication.

When purchasing any kind of propagator always make sure it has sufficient heating capacity for your requirements. A propagator for a cold greenhouse will need to have more in reserve than one in an already heated environment. With thermostatic control a greater heating capacity will cost no more to run than a less powerful unit. As a rule I would always advise buying the

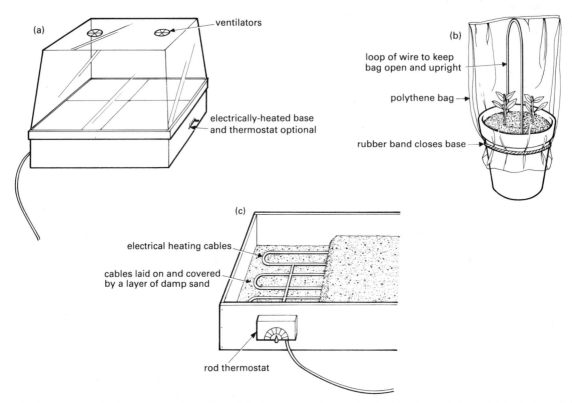

Fig. 43 Propagators. (a) A commercially-produced electric propagator. (b) A polythene bag propagator. (c) A home-made propagator base.

largest and best you can afford. Propagators, like most gardens, never seem large enough.

A note might be made here regarding mist propagation. This has led to a great improvement in the rooting potential of many previously 'difficult' cuttings, most notably among conifers and other evergreens. A mist unit is based on a specially constructed propagating frame containing under-bench heating controlled at round 21 °C and mist irrigation (Fig. 44).

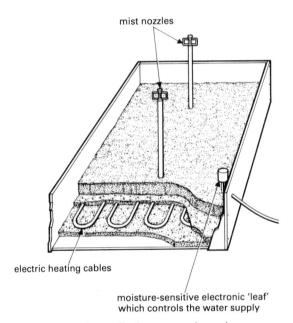

mist nozzles

electric heating cables

moisture-sensitive electronic 'leaf'
which controls the water supply

Fig. 44 Layout of a small mist propagation unit.

The mist is provided by fine nozzles set above the bed and these keep the plants continuously moist and the atmosphere around them at maximum humidity. The amount of mist is controlled by an electronic 'leaf' which senses when the atmosphere becomes dry and operates a solenoid valve which in turn turns on the mist until the necessary dampness is achieved.

Owing to the continuous warm, damp atmosphere, cuttings under mist do not wilt and require no shading.

Under these conditions they continue to grow and soon form roots.

A free-draining compost is required, this usually consisting of silver sand alone or mixed with a small amount of peat. Cuttings are either rooted in this compost in a bed or in boxes which ideally are fitted with mesh bottoms to assist with drainage. There are no nutrients in the compost, and therefore plants rooted in this way must be transplanted as soon as roots are established. After transplanting they should be kept in a shaded, close atmosphere for a time, gradually weaning them to an ordinary greenhouse environment.

Amateur units are available in kit form, or individual components can be bought.

The technique used in mist propagation is not new but merely mechanised. Nevertheless it can be duplicated by the dedicated gardener with a frame with bottom heat, the dampness being supplied by means of a syringe which is applied frequently in sunny weather and not so much in dull. Constant attention is necessary so it is not a method for those out at work all day.

All greenhouses, frames and propagators, especially those creating a warm damp atmosphere, should be kept as hygienic as possible by washing them down with disinfectant whenever possible. All disease and pests which do appear should be dealt with immediately one way or another.

Lastly we go back to the beginning with a house and a garden. Without any other facilities there is always the kitchen windowsill. A propagator of modest size does not look out of place there and neither does the home-made polythene bag propagator. This consists simply of a pot of compost in which cuttings or seeds are placed and a polythene bag cover which retains the necessary warm, close atmosphere provided by its larger cousins. The bag is simply popped over the top of the pot and held in place with a rubber band. Care should be taken to keep it away from any leaves by supporting it with either three short canes or a loop of stiff wire placed inside the pot. Some ventilation is necessary, especially when cuttings are just beginning to root, and is simply provided by snipping away the corners of the bag.

This method can also be used outside, preferably placing the pot with its cover in gentle shade under a bush or similar position until its mission is accomplished.

Part 2
PROPAGATION
of INDIVIDUAL PLANTS

In the following chapters will be found details of the propagation requirements of many cultivated plants. These can be located within each chapter under their generic Latin name, or, where they have common names, by reference to the index. To list every species would be virtually impossible, and in most cases unnecessary, so the information is given for each genus as a whole. Particular species are mentioned only if their requirements differ from the norm.

Natural layering of
stems (*Sedum* species)

Erinus alpinus (seedling)

Rooted offset (*Sempervivum*)

if birds or vermin prove a problem,
cover pots with a sheet of perforated
zinc, wire gauze or fine wire netting

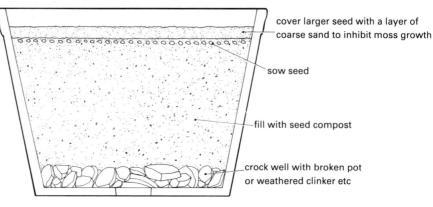

cover larger seed with a layer of
coarse sand to inhibit moss growth

sow seed

fill with seed compost

crock well with broken pot
or weathered clinker etc

sowing alpines to stand outdoors to stratify

Draba aizoides
(seedling and root system)

Lithospermum diffusum
(heel cutting)

Armeria maritima
(seedhead and seeds)

Mossy saxifrage (seedling)

8 Alpines and Rock Garden Plants

This chapter includes the small plants most commonly used in a rock garden or scree, but excludes dwarf bulbs and shrubs as these will be found in the relevant group. This principle has been followed throughout this part of the book.

Discounting these latter examples still leaves a vast range of plants with a diversity of cultural requirements almost equalling their numbers. A great proportion, however, will grow happily in the sunny rock garden in any good, well-drained fertile soil. For those that will not, the following brief notes on composts and special conditions may prove useful.

Winter damp rather than cold is the enemy of many alpines, and overhead protection in the form of a sheet of glass or similar is useful for susceptible species. Alternatively, such plants can be grown in pans throughout the summer in a plunge bed and be housed in a special greenhouse or frame in the winter. Such a greenhouse is termed an 'alpine house', and is basically constructed to allow the maximum admission of light and ventilation. Most are cold houses, ventilation being given at all times regardless of temperature, the house only being closed during storms or damp foggy weather.

A suitable compost for pans is one consisting of 2 parts (by volume) coarse sand, 2 parts leaf-mould or peat and 1 part loam with a sprinkling of bonemeal. For plants disliking lime a mixture of 1 part loam, 2 parts moss peat and 1 part coarse sand should suit. This, incidentally, is suitable for most lime-hating plants.

Alpines are often grown in troughs or sinks and a compost of equal quantities by volume of loam, moss-peat or leaf-mould and coarse sand is suitable for this purpose, again with a small amount of bonemeal added.

The terms 'moraine' and 'scree' are used occasionally, these being very similar horticulturally. Both are formed from rock rubble and stones in which a little compost has been incorporated. They are usually formed in special beds, although deep containers can also be used. Scree is the more open of the two, and both are surfaced with a thick layer of limestone or granite chippings. Ideally moraine should also be supplied with water which will run through its base during the growing season.

For seed sowing the J.I. Seed Compost suits most types, the same being used for pricking out and potting on. Alpine seedlings are invariably very small so they should be handled with care, the smallest being grown on in frames until well established. Perfect drainage is essential in most cases, so crock all pots well.

When stratifying seeds place out in a northerly aspect, if possible, away from the drip of trees or buildings, and never let the compost dry out.

PLANT NAME	METHOD OF PROPAGATION	TREATMENT REQUIRED	TIME OF YEAR PROPAGATED	CULTURAL NOTES
Acaena (New Zealand Burr)	DIVISION	Replant divisions in well-drained soil. Allow for spread of 45 cm.	Spring or autumn	Evergreen carpeting plants for sun or shade. *A. microphylla* will grow beneath conifers.
	SEED	Sow in cold frame. Plant seedlings out in autumn.	March to May	
Acantholimon (Sea Lavender)	SEED (slow to germinate)	Sow in warm shady spot in greenhouse or frame.	As soon as ripe or late spring	Evergreen cushion forming plants for warm sunny position in sandy well-drained soil. Overwinter cuttings and seedlings in frames.
	LAYERING	Peg outlying shoots down to root and separate in spring.	During summer	
	SEMI-HARDWOOD HEEL CUTTINGS	Root in sandy compost in cold frame or propagator.	September	
Actinella (A. grandiflora, syn. Rydbergia grandiflora)	DIVISION	Replant divisions 7–10 cm apart in sandy well-drained soil in full sun.	March	Short-lived perennials for alpine house or rock gardens in dry areas. Can also be raised from seed when available.
Aethionema (Burnt Candytuft/Candy Mustard)	SEED	Sow annual types where they will flower, perennials in pans of seed compost mixed with old mortar rubble in cold frame. Overwinter in frame.	When ripe or in spring	Annuals and perennial sub-shrubs. Require well-drained spot in full sun. Perennials flower second year from spring-sown seed. Propagate *A. coridifolium* 'Warley Rose' from cuttings only.
	SEMI-HARDWOOD HEEL CUTTINGS	Root in cold frame and overwinter there. Plant out in spring allowing for 20–25 cm spread.	Summer, after flowering	
Alyssum (Hardy perennial types; A. saxatile = Gold Dust)	SEED (stratify high alpine types)	Sow in cold frame or greenhouse. Requires stratification. Plant out in autumn or earlier if ready.	During spring or early summer	Most species ideal for sunny rock garden or edging. Taller types are border subjects. Plant out 23–30 cm apart.
	SOFTWOOD STEM CUTTINGS	Root in shady place outdoors. Plant out in autumn or pot and overwinter in frame.	July to August	
Anagallis (Pimpernel; A. arvensis = Poor Man's Weather Glass)	SEED	Sow *A. caerulea* in cold frame and harden off before planting out. Sow others where they will flower.	March/April	Plant in sunny spot. Best treated as annuals though some are perennial. Boggy conditions suit *A. tenella*.
Androsace (Rock Jasmine)	SEED (annual types grown from seed only)	Sow in pots and place outdoors to stratify. Bring into cold frame or greenhouse in March. Grow on in frames until established.	As soon as ripe or December/January	Many species ideal for sinks, troughs or alpine house spreading from 8 to 30 cm depending on species. Some types produce runners which can be layered in the same way as strawberries. Grow on as for divisions.
	DIVISION	Remove small rosette offsets from species on which they appear and pot up. Overwinter under glass and plant out when large enough.	August	
Andryala	SEED	Sow in cold frame or greenhouse and grow on in frame until established.	As soon as ripe or in spring	Herbaceous perennials grown for foliage as well as flowers. Suitable for alpine house or well-drained rock garden. Can also be divided in spring.
	HEEL CUTTINGS	Root in cold frame and overwinter there.	Late summer	
Antennaria (Mountain Everlasting)	SEED	Sow in pots and place out to stratify. Move to a cold frame in March.	January/February	Spreading mat-forming plants with silver/grey leaves for a well-drained position in rock gardens or as edging.
	DIVISION	Replant in the open ground.	March/April	

PLANT NAME	METHOD OF PROPAGATION	TREATMENT REQUIRED	TIME OF YEAR PROPAGATED	CULTURAL NOTES
Anthyllis (Kidney Vetch; A. vulneraria = Ladies' Fingers)	HEEL CUTTINGS (shrubby types)	Root in shaded cold frame and grow on there. Plant out late next spring.	June to August	Mostly low-growing shrubs and herbaceous perennials requiring warm sunny spot in well-drained soil. Some species are tender and need protection in severe winters. A. vulneraria, a native species, is probably the hardiest.
	DIVISION (herbaceous perennials)	Replant divisions where they will flower.	In spring	
	SEED	Sow in cold frame or greenhouse and grow on there until established. Plant out in spring or autumn.	In spring	
Arabis (Rock Cress)	SOFTWOOD CUTTINGS	Root in a shady spot outdoors or in cold frame. Overwinter in frames.	During summer	Ideal plants for edgings, walls and banks, spreading from 30 to 45 cm.
	SEED	Sow in cold frame or prepared bed outdoors. Prick out or thin and plant out in spring.	July	
	DIVISION	Replant divisions where they will flower.	Spring or autumn	
Arenaria (Sandwort)	SOFTWOOD CUTTINGS	Root small cuttings in sandy compost in a cold frame and overwinter there.	During summer	Low-growing or creeping plants. Most need a sunny spot in well-drained but moist soil. Grow A. balearica among paving or in shade of rocks. Plant out allowing for spread of 30–38 cm in most cases.
	DIVISION	Separate rooted pieces from around older plants.	April	
	SEED	Sow in cold frame or prepared bed outside. Thin or prick out and plant out in autumn.	During spring	
Arisarum (A. proboscideum is hardy; others require greenhouse treatment)	DIVISION (of tuber)	Lift and divide rhizome-like roots of established clumps. Replant 15 cm deep.	In spring	Low-growing herbaceous plants with flowers like small long-tailed mice. A semi-shaded spot in humus-rich soil suits it best. Can also be raised from seed when available.
Armeria (Thrift/Sea Pink)	DIVISION	Carefully remove portions from edges of older plants. Replant in open ground 23–25 cm apart.	In spring or autumn	Ideal for sunny rockery or as edging. Seed of species such as A. caespitosa require stratification for best germination. Plant seedlings in final positions in autumn or spring.
	SEED	Sow in cold frame. Prick out into nursery beds outdoors or boxes in frames.	Late spring or early summer	
Asperula (Woodruff)	SEED	Sow in cold frame and grow on there. Plant out autumn or spring. Sow annuals where they will flower and thin to 23 cm apart.	March/April	Mostly herbaceous plants for sunny positions or alpine house. A. odorata, a native species, is best in a shady spot. This plant smells of new mown hay and when dried and crushed can be used among clothes in place of mothballs.
	DIVISION	Replant in ordinary soil allowing for a 23–25 cm spread in most cases.	In spring or early summer	
Aubrieta or Aubretia (Rock Cress)	SEED	Sow in greenhouse or propagator at 18–24°C or in prepared bed outdoors a little later.	March	Evergreen trailing plants for the rock garden or walls. Trim right back with shears after flowering. Seed will not breed true so raise selected strains vegetatively. Plant out overwintered plants in spring 25–30 cm apart.
	SOFTWOOD CUTTINGS	Root in shady spot outdoors or in cold frame. Protect over winter with frame or cloche.	September/October	
	LAYERING	Peg shoots down and separate in autumn when roots have formed. Pot and overwinter in frame.	During summer	
	DIVISION	Replant divisions where they will flower.	In autumn	

PLANT NAME	METHOD OF PROPAGATION	TREATMENT REQUIRED	TIME OF YEAR PROPAGATED	CULTURAL NOTES
Calamintha *(Calamint)*	SEED	Sow in pots and place outside to stratify. Move to cold frame or greenhouse in March. Grow on in cold frame.	January	Sub-shrubby plants with flowers rather like thyme. Require a moist soil in sun or semi-shade. In all cases plant out in final positions in autumn 20–25 cm apart.
	BASAL CUTTINGS	Root in greenhouse or cold frame.	May	
Celmisia *(C. coriaiea = Snow Daisy)*	DIVISION	Carefully remove sections from established plants. Pot individually and grow on in cold frame.	During spring	Difficult but desirable tufted plants from New Zealand. Most require a well-drained yet moist peaty soil in full sun, or scree conditions. Overhead protection in winter is necessary in most districts. Can be raised from seed, but this is rarely available.
	BASAL CUTTINGS	Cuttings can be made from side rosettes, rooting them in well-drained compost in cold frame. Grow on there until established.	During summer	
Cerastium *(Snow In Summer, a name sometimes applied to Arabis caucasida, syn. A. albida)*	SEED	Sow in cold frame or prepared bed outside. Prick out or thin and plant out in autumn.	As soon as ripe or in spring	Annuals and perennials useful as ground cover. Some are inclined to be weedy. *C. alpinum* is a native species in cultivation being a good alpine or house plant. Choicer types may need overhead winter protection.
	SOFTWOOD CUTTINGS	Root in a cold frame or shady spot outdoors.	During summer	
	DIVISION	Replant pieces from the edges of older plants according to spread of the species.	September	
Chiastophyllum *(syn. Cotyledon oppositifolia)*	SOFTWOOD CUTTINGS	Root in frame or greenhouse. Grow on in pots and overwinter in frame.	July to September	Hardy succulent ideal for rock garden or dry walls. Requires a sunny well-drained position. Prick out seedlings in frames and harden off before planting out.
	SEED	Sow in a warm greenhouse or propagator at 15–18°C.	February/March	
Chimaphila *(C. umbellata = Prince's Pine)*	SEED (minute)	Sow in peat-based compost in shady cold frame or greenhouse.	March to May	Difficult evergreen sub-shrubs which require a shady spot in moist but well-drained lime-free soil. Grow seedlings on in a similar compost and protect in a frame over winter.
	DIVISION	Bits can be carefully separated from the creeping rootstock of older plants. Grow on in pots until well rooted.	In spring	
Chrysogonum	DIVISION	Replant divisions from older plants 23 cm apart in peaty lime-free soil or compost.	In spring	A long-lived herbaceous perennial for a sunny or semi-shaded position.
Conandron	DIVISION (tuberous rooted)	Divide established tubers and repot separately. Grow on in frame or greenhouse.	Autumn or early spring while plant is dormant	A plant related to *Streptocarpus*. Requires a well-drained acid soil with much leaf-mould. Grow in alpine house in all but the warmest, driest areas. Seedlings are small, so grow on in frame or greenhouse.
	SEED (dust-like)	Sow in a sand/peat compost in slight warmth under glass just pressing seeds into compost surface.	March/April	
Convolvulus *(C. arvensis is the Bindweed so difficult to eradicate from gardens)*	DIVISION (perennial types)	Replant divisions according to species either in rock garden or pots of humus-rich compost for growing on under glass.	In spring	Mostly colourful twining plants, a few being ideal for sunny rock garden or alpine house. Some species require greenhouse conditions throughout their lives. Grow seedlings on in cold frame until established.
	SOFTWOOD CUTTINGS (hardy perennials)	Root in propagator or cold frame and overwinter in frame.	August	
	SEED	Sow hardy annuals where they will flower, the rest in propagator or cold frame.	March/April	

PLANT NAME	METHOD OF PROPAGATION	TREATMENT REQUIRED	TIME OF YEAR PROPAGATED	CULTURAL NOTES
Coptis (Goldthread/Mouth Root)	DIVISION	Replant divisions in flowering position 10–12 cm apart.	Early spring	Hardy evergreen plants resembling wood anemones. Require a moist peaty soil in semi-shade. A plant for that boggy corner. Plant out in spring or autumn.
	SEED	Sow in shady cold frame and grow on there. Overwinter under glass if necessary.	As soon as ripe or in spring	
Coronilla (C. varia = Crown Vetch; C. emerus = Scorpion Senna)	SEED	Sow hardy types in pots and stand outside to stratify. Transfer to warm propagator in March. Tender species are best sown in heat at 18–24°C.	December/January March/April	Shrubs, sub-shrubs and perennials, some suited to rock garden and alpine house. A few such as C. cappadocica are hardy, this being a trailing plant. Grow seedlings on under glass to be potted or planted out according to species.
	SEMI-HARDWOOD CUTTINGS (shrubby types)	Root in cold frame or greenhouse and overwinter there.	August/September	
	LAYERING	Peg down shoots and separate when rooted. Transplant to final positions in spring.	During summer	
Cortusa	SEED	Sow in shaded cold frame. Grow on there and plant out early in autumn.	As soon as ripe or in spring	Woodland plants requiring a moist, leafy type of soil in shade or semi-shade. Plant out 15–23 cm apart.
	DIVISION	Replant divisions of old clumps where they will flower.	After flowering	
Corydalis (Fumitory)	DIVISION (tubers and offsets)	Replant divisions in flowering positions giving space according to species.	In spring	Plants for semi-shade in moist but well-drained soil. Many suitable for the rock garden. C. lutea is a native species in cultivation. Taller types are best grown in woodland conditions.
	SEED	Sow annuals in their flowering positions, perennials in cold frames or prepared beds outdoors. Thin or prick out and plant in final positions in autumn.	March/April	
Cotula (C. coronopifolia = Brass Buttons)	DIVISION	Replant rooted pieces from established plants 45 cm apart.	In late summer and autumn	Carpeting plants for a sunny position. Make a good foil for dwarf bulbs. Use C. squalida in cracks between paving.
Cyananthus	SEED	Sow in cold frame and prick out 1 cm apart. Grow on in pots until planting out in autumn.	March to May	Low-growing plants for moist, humus-rich but well-drained soil. Grow in rock garden or scree conditions. Some species are herbaceous, dying back for the winter to reappear in spring. Division is rarely successful.
	SOFTWOOD CUTTINGS	Use non-flowering shoots and root in a sand/peat compost in a cold frame. Overwinter in frames.	Early summer	
Douglasia	SEED	Sow in pots and place outside to stratify. Transfer to slight heat in March. Grow on under glass through the winter.	As soon as ripe or in December/January	Evergreen plants suitable to alpine house or trough culture. Require a compost of peat and leaf-mould with enough sand and grit incorporated to give perfect drainage.
Draba (D. aizoides, Yellow Whitlow Grass, is a native species which is reasonably winter-resistant)	SEED	Sow thinly in pots and expose to frost and snow. Bring into slight warmth in March. Grow on in pots during summer outdoors and overwinter in cold frame.	As soon as ripe or in December/January	Mostly tufted or cushion-forming plants. Many require alpine house protection in all but the driest areas. Annual types will behave as biennials if sown in autumn where they are to flower.
	DIVISION (many perennial types)	Remove side rosettes and pot singly, placing them in a plunge bed outdoors to be brought under glass for the winter.	August	

PLANT NAME	METHOD OF PROPAGATION	TREATMENT REQUIRED	TIME OF YEAR PROPAGATED	CULTURAL NOTES
Dryas *(D. octopetala =* *Mountain Avens, a native* *species growing in* *limestone areas)*	SEED	Sow in pots outdoors to stratify. Transfer to slight heat in March.	December/January	Long-lived evergreen sub-shrubs for a sunny situation in ordinary soil. Roots are damaged easily, so transplant carefully into individual pots. Seedlings take 2–3 years to reach flowering size.
	HEEL CUTTINGS	Root in shaded cold frame or under cloches. Overwinter under glass. Plant out 30–35 cm apart in spring.	August	
	LAYERING	Peg shoots down and separate carefully. Pot individually and overwinter in frame.	After flowering	
Edraianthus	SEED (seedlings generally variable)	Sow in cold frame and grow on there or in plunge bed outside until following spring.	February/March	Low-growing tufted plants for a deep, well-drained soil in sun. Mix a quantity of limestone chippings in final planting position. Plant out 12–20 cm apart.
	BASAL CUTTINGS	Root 4–5 cm cuttings in a peat/sand compost in cold frame, overwintering there.	July/August	
Empetrum *(E. nigrum = Crowberry)*	SEMI-HARDWOOD CUTTINGS	Root in sand/peat compost in cold frame. Overwinter there and plant out in spring.	During summer	Low-growing evergreen shrubs useful as ground cover on moist lime-free soils.
Epigaea *(E. repens = May* *Flower/Trailing Arbutus)*	SEED	Sow in pots outdoors to stratify. Bring into slight heat in March. Grow on in individual pots.	As soon as ripe	Low-growing or trailing evergreen shrubs for acid, humus-rich soil in semi-shade. Some species need overhead winter protection. Overwinter all young plants in cold frames. *E. repens* forms layers naturally.
	SEMI-HARDWOOD CUTTINGS	Root short cuttings in sand/peat compost in shaded cold frame.	During summer	
	LAYERING	Peg stems down and separate when rooted. Pot up individually.	During summer	
Erinus *(E. alpinus = Summer* *Starwort)*	SEED (will self-seed when established)	Sow in cold frame or greenhouse. Grow on in boxes or nursery bed outdoors.	In spring	A short-lived perennial and ideal for crevices in rocks and paving. Plant seedlings out in autumn.
	DIVISION	Replant divisions in any suitable spot.	At any time	
Eriogonum *(E. fasciculatum =* *Californian Buckwheat)*	SEED	Sow under glass in slight heat. Grow on in frames.	February to April	Genus contains many perennials and sub-shrubs for the rock garden. Require a rich peaty soil in full sun. Woolly or hairy-leaved species may need overhead winter protection. Plant out young plants in late spring.
	DIVISION	Replant divisions where they will flower or grow on in cold frame.	March	
	SEMI-HARDWOOD HEEL CUTTINGS (shrubby types)	Root in sand/peat compost in cold frame and overwinter there.	July to September	
Eriophyllum *(E. lanatum, syn. Bahia* *lanata = Oregon* *Sunshine)*	SEED	Sow in cold frame or prepared bed outdoors. Plant out in autumn.	April	Perennial plants for rock garden or border. Need a warm, well-drained position in sun.
	DIVISION	Replant divisions where they will flower.	March or before growth ends in autumn	
Erysimum *(Fairy Wallflower/Hedge* *Mustard)*	SEED	Sow annuals where they will flower, the remainder in prepared beds outdoors planting in final positions in August.	April June	Annuals, biennials and perennials related to wallflowers. Many are good rock garden plants. *E. pumilum* suitable for troughs. Most grow in ordinary soil in sun.
	DIVISION (perennials)	Replant divisions where they will flower.	March/April	
	HEEL CUTTINGS (perennials)	Root in cold frame and overwinter there.	July/August	

PLANT NAME	METHOD OF PROPAGATION	TREATMENT REQUIRED	TIME OF YEAR PROPAGATED	CULTURAL NOTES
Frankenia *(F. laevis = Sea Heath)*	DIVISION	Replant divisions in well-drained sandy soil 10–12 cm apart.	September to April	Evergreen carpeting plants with creeping stems. Several species suitable for troughs and sinks or the rock garden in a sunny spot.
	SEED	Sow in cold frame and grow on there until large enough to plant out.	As soon as ripe or in March	
Gentiana *(Gentian; G. verna = Spring Gentian)*	SEED	Sow in a gritty compost and place outdoors to stratify. Bring into slight heat in March. Leave smallest seedlings in their seed pans for a year before moving.	As soon as ripe or in December/January	Genus contains many species for almost every situation in the garden. Many are difficult subjects. Many rock garden types require a fairly sunny spot in well-drained yet always moist soil.
	DIVISION	*G. acaulis* and those with similar habits can be carefully divided.	Early in spring	
	BASAL CUTTINGS	*G. septemfida* and *G. ornata* can be raised from shoots 2.5–5 cm long rooted in shaded greenhouse or cold frame.	In spring when basal shoots are long enough	When rooted pot singly and grow on in shaded frames. Cuttings can be made from other species, especially those not coming true from seed.
Globularia *(G. vulgaris = Blue Daisy)*	SEED	Sow in cold frame on compost surface pressing in slightly. Grow on in frames.	March/April	Long-lived plants for dry sunny spots especially in limy areas. Overwinter seedlings and cuttings under glass planting out in spring 10–20 cm apart depending on species.
	SEMI-HARDWOOD CUTTINGS (shrubby types)	Root in cold frame or under cloches.	During summer	
	DIVISION	Replant divisions where they will flower.	October to April	
Haberlea	SEED	Sow in cold frame or greenhouse and grow on in pots.	When ripe or March/April	Shade-loving plants growing well in crevices in a north-facing wall in moist rich soil with plenty of leaf-mould. Seedlings and cuttings are best grown on and overwintered in frames.
	LEAF CUTTINGS	Use fully grown undamaged leaves complete with stalk. Plant in a sand/peat compost in a shaded greenhouse. Pot singly when small plants have formed.	During summer	
Hacquetia	DIVISION	Divide strong plants carefully before new growth begins and replant in semi-shade in good soil 10–12 cm apart.	In spring	An herbaceous perennial which resents disturbance. Will self-seed and stock can be increased by potting up seedlings which appear around older plants.
Helianthemum *(Rock Rose)*	SEED	Sow shallowly in a prepared bed outdoors or in a cold frame during February and March. Plant out in autumn.	Late spring and early summer	Flowers resemble real roses though they are unrelated. They prefer a sunny spot in open soil preferably over limestone or chalk. Autumn-rooted cuttings are best overwintered under glass planting out in spring 20–30 cm apart.
	TIP CUTTINGS	Root non-flowering shoots in shaded cold frame or cloche and pot singly when rooted.	Early spring or in autumn	
Helichrysum *(Everlasting Flower)*	SEED	Sow annuals in slight heat under glass and harden off before planting out 23–30 cm apart in May. Sow perennials in open ground or in cold frame.	February April to July	Mostly plants for the alpine house or well-drained spot in sun in drier areas. Many species are not hardy and some are used as half-hardy annuals for bedding out.
	HEEL CUTTINGS (perennial and shrubby types)	Root 7.5 cm cuttings in cold frame or greenhouse. Grow on under glass.	During summer	
Hepatica	SEED	Sow in a shady bed outdoors. Thin slightly and plant out in autumn or spring.	As soon as ripe or in spring	Close relations of the anemones and often on sale as that genus. They require semi-shaded woodland conditions on lime rich soils. Plant out 12–15 cm apart.
	DIVISION	Replant divisions where they will flower.	In autumn when foliage dies down	

PLANT NAME	METHOD OF PROPAGATION	TREATMENT REQUIRED	TIME OF YEAR PROPAGATED	CULTURAL NOTES
Houstonia (H. caerulea = Bluets)	DIVISION	Replant divisions 30 cm apart in a moist lime-free soil in a north-facing position.	March/April	Over-enthusiastic flower-producers which may flower themselves to death. *H. caerulea* is best grown as an annual or biennial. Ideal plants for the alpine house.
	SEED	Sow shallowly in a compost of leaf-mould and sand in a cold frame.	As soon as ripe or in March	
Hutchinsia	DIVISION	Divide carefully and pot singly to grow on in cold frame.	March/April	Small plants, many suitable for troughs, sinks or an open well-drained place in rock garden. The annual *H. stylosa* will self-seed.
	SEED	Sow shallowly where plants are to flower.	March/April	
Hypsela	DIVISION (stems root at nodes as they creep)	Replant divisions to allow for a spread of around 30 cm.	March	Carpeting plants for a moist semi-shaded spot in peaty soil.
Iberis (Perennial Candytuft)	HEEL CUTTINGS	Root in cold frame or shady spot outdoors. Give glass protection over winter.	During summer	Evergreen plants ideal for the tops of walls where they are well-drained and in sun. Trim back with shears after flowering. Plant out young plants in spring.
	DIVISION	Replant divisions where they will flower 25 cm apart.	In spring	
	SEED	Sow in cold frame.	April	
Jasione (J. montana = Sheeps Bit Scabious)	SEED	Sow annuals where they will flower. Sow perennials in a prepared bed outside.	March In autumn	Annuals and herbaceous perennials, most suited to the sunny rock garden. Plant seedlings in final positions in spring or autumn.
	DIVISION	Replant divisions where they will flower.	March	
Jeffersonia (some species also found under Plagiorhegma)	DIVISION	Replant divisions where they will flower 10–15 cm apart.	In spring as growth begins	Herbaceous woodland plants for semi-shade. Rock garden or alpine house in a leaf-mould based compost. Seedlings may take 3 to 5 years to reach flowering size.
	SEED (may take a year or more to germinate)	Sow in cold frame and leave seedlings undisturbed for one year. Grow on in frames.	As soon as ripe	
Kelseya	SEED	Sow in cold frame and grow on there.	As soon as ripe	A cushion-forming plant for the alpine house in a gritty compost in sun.
	HEEL CUTTINGS	Root small sideshoots in cold frame or greenhouse.	Late spring or early summer	
	DIVISION	Replant in pans for the alpine house.	In spring	
Leontopodium (L. alpinum = Edelweiss)	SEED	Sow in cold frame. Prick out into gritty compost and overwinter in cold frame.	March	Requires a light soil in sunny spot or alpine house. Harden off before planting out in May 15–23 cm apart. Divided plants are difficult to re-establish.
	DIVISION	Divide carefully to avoid damaging roots. Replant where they will flower.	In spring	
Leucogenes (L. leontopodium syn. Helichrysum leontopodium)	SEED	Sow in cold frame and grow on there until established.	April	Evergreen sub-shrubs, the New Zealand equivalent of edelweiss. Grow outdoors in warm areas only, elsewhere in alpine house. Prefer a semi-shaded position.
	HEEL CUTTINGS	Make 2.5 cm cuttings of non-flowering shoots and root and grow on in cold frame.	Early summer	
Lewisia	SEED (variable results as they freely hybridise)	Sow in cold frame growing on and overwintering there.	March	Evergreen and herbaceous plants for full sun in a rich lime-free soil. Give overhead winter protection from wet. Plant out overwintered plants in April 15–23 cm apart.
	DIVISION (offsets)	Replant offsets in individual pots and grow on and overwinter in frames.	June	

PLANT NAME	METHOD OF PROPAGATION	TREATMENT REQUIRED	TIME OF YEAR PROPAGATED	CULTURAL NOTES
Linnaea (L. borealis = Twinflower)	DIVISION	Replant divisions where they will grow 30–45 cm apart.	In spring	Evergreen creeping sub-shrub for a moist semi-shaded position. Makes good ground cover. Can also be propagated from cuttings which root slowly.
	LAYERING	Peg down non-flowering shoots and separate in spring.	During summer	
Lithospermum (L. purpurocoeruleum = Purple Gromwell/Walking- stick Plant)	HEEL CUTTINGS (L. diffusum)	Root in propagator or greenhouse and overwinter in cold frame.	July/August	Evergreen sub-shrubs for sunny positions. L. diffusum (e.g. Lithospermum 'Heavenly Blue') is prostrate spreading about 30 cm in acid sand/leaf-mould soil. Others like lime.
	DIVISION (L. graminifolium, syn. Moltkia suffruticosa)	Replant in open ground 18–20 cm apart.	September or March	
Mazus	DIVISION	Replant divisions 30 cm apart where they will flower.	Autumn or spring	Low-growing creeping plants similar to mimulus. Need a sunny well-drained spot. Grow well in pans for alpine house.
	SEED	Sow in cold frame and grow on there until following spring.	April	
Micromeria	DIVISION	Replant divisions 23–25 cm apart in open ground.	March/April	Long-lived sub-shrubs for sunny well-drained positions. Most make neat mounds of foliage. M. chamissonis is raised from stolons.
	STEM CUTTINGS	Take small cuttings of non-flowering shoots and root in cold frame. Plant out in spring.	May/June	
Mimulus (Musk/Monkey Flower)	TIP CUTTINGS	Root in cold frame or under cloches in moist soil and overwinter there.	August	A plant for moist sunny spot though M. primuloides will grow in shade. Harden off seedlings before planting out in May. M. glutinosus is a shrub for a cool greenhouse.
	DIVISION	Taller species can be divided, replanting 25–30 cm apart.	In spring or early summer	
	SEED	Sow in greenhouse or cold frame.	March	
Minuarta (Genus includes the Sandworts)	SEED	Sow in cold frame and grow on in small pots until large enough to go out.	During spring	Prostrate or cushion-forming plants for trough or sink culture. Often wrongly offered as Arenaria or Alsine.
	DIVISION	Separate rooted pieces from around older plants.	April	
Mitchella (M. repens = Partridge Berry)	DIVISION	Trailing stems root at the nodes and can be separated and replanted.	In early spring	Evergreen trailing plants for semi-shade in a peaty soil or for pans in alpine house.
Mitella (Mitrewort/Bishop's Cap)	SEED	Sow in cold frame in moist, peat-based compost and grow on in small pots.	March to May	Perennial woodland plants for semi-shaded position in moist sand/leaf-mould soil or mossy crevices in rock garden.
	DIVISION	Replant in growing position.	March/April	
Moltkia (see also Lithospermum)	HEEL CUTTINGS	Root in propagator or greenhouse and overwinter in frame.	July/August	Sub-shrubby plants related to Lithospermum and requiring a sunny position in good soil with lime or a pan in the alpine house.
	LAYERING	Peg lower shoots down and separate when rooted.	During summer	
Morisia (Monotypic. M. monantha, syn. M. hypogaea = Mediterranean Cress)	ROOT CUTTINGS	Cut sections 2.5 cm long and root in moist sand in cold frame. Pot up when rosettes have formed.	February/March	A good plant for moraine, sinks and troughs, or alpine house in gritty compost. Plant out in September 15 cm apart. Seed capsule buried naturally by drooping of flowers after blooming.
	SEED	Sow in cold frame and grow on there.	March/April	

PLANT NAME	METHOD OF PROPAGATION	TREATMENT REQUIRED	TIME OF YEAR PROPAGATED	CULTURAL NOTES
Myosotis *(Forget-me-not)*	SEED	Sow in prepared bed outdoors or in cold frame for smaller species.	During early summer	Many species other than those used as biennial bedding plants, a number are suitable for rock, trough or sink gardens and the alpine house. Most prefer a moist soil in full sun. Plant out in spring.
	DIVISION (perennials)	Replant according to species' requirements	Early autumn or spring	
	SOFTWOOD CUTTINGS	Root cuttings from non-flowering shoots in shady place outdoors or in cold frame.	During summer	
Nierembergia *(Cup Flower)*	DIVISION	Replant in sun or semi-shade in moist soil 30 cm apart.	In spring as growth begins	*N. repens* is a mat-forming plant.
Onosma *(O. tauricum = Golden Drop)*	STEM CUTTINGS (perennials)	Root in shaded propagator and avoid excessive damp on foliage. Grow on in frames.	July	Hairy-leaved plants requiring a hot, well-drained position in full sun or alpine house. Seed of some hardy species may need stratification.
	SEED	Sow annuals where they will flower, perennials in cold frame. Grow on there.	April March/April	
Orchids *(hardy species suitable to rock garden or alpine house culture will be found among the following genera: Cypripedium = Slipper Orchid; Goodyera repens = Adder's Tongue; Spiranthes = Ladies' Tresses)*	DIVISION (rhizomatous or tuberous roots etc) DIVISION (offsets)	Divide *Cypripediums* when repotting. Lift and carefully divide older groups of *Gymnadenia* (slow to multiply), *Epipactis* (creeping rootstock) and *Cephalanthera*. Remove offsets from *Bletilla* and *Pleione* when repotting, and from established colonies of *Calypso*, *Ophrys*, *Orchis* and *Spiranthes*.	In spring as growth begins In autumn In spring In autumn	Generalised cultural notes: *Bletilla*, *Cypripedium* and *Pleione* alpine house in compost 4 parts fibrous loam, 1 sand, 1 leaf-mould and 1 finely chopped sphagnum moss. Outdoors woodland type soil in semi-shade suits most *Calypso*, *Cephalanthera* (with lime), *Epipactis* and *Goodyera repens*. Deep sandy loam with underlying chalk or limestone in sun suits *Ophrys*, *Orchis* and *Spiranthes*.
Oreocharis	SEED	Sow in slight heat under glass and grow on in frames.	February to April	Rosette-forming alpine house plants requiring a lime-free leafy soil with good drainage.
	LEAF CUTTINGS	Use fully grown leaf with stalk. Root in sand/peat compost in shady spot in greenhouse.	During summer	
Orphanidesia	SEED	Sow in pots outdoors to stratify. Transfer to slight heat in March. Grow on in small pots.	December/January	An evergreen mat-forming shrub for a rich, acid woodland soil in semi-shade. Plant out in spring.
	SEMI-HARDWOOD CUTTINGS	Root cuttings in sand/peat compost in shaded cold frame.	During summer	
Ourisia	SEED	Sow in cold frame and grow on overwintering there.	April	Rather tender plants for well-drained position out of the full midsummer sun where they make good carpeting plants.
	DIVISION	Replant divisions 38–46 cm apart.	April	
Oxytropis	SEED	Sow in prepared soil where they will flower.	April/May	Many dwarf perennials preferring a light sandy soil in good light.
Paraquilegia *(P. grandiflorum, syn. Isopyrum grandiflorum)*	DIVISION	Carefully divide older plants and replant as grown. Can also be grown from seed where available	In spring as growth begins	Delicate tufted plants for alpine house or cool crevices in rock gardens in north. Scree conditions.
Parnassia *(P. palustris = Common Grass of Parnassus)*	DIVISION	Divide and replant 15 cm apart in appropriate place.	March/April	Plants for a moist shady position in peat or bog conditions.
	SEED	Sow in moist peat in a shady spot outdoors.	As soon as ripe or in spring	
Parochetus	DIVISION (natural layers)	Stems root as they advance during growing season. Separate and grow on under glass planting out in spring.	Late summer/early autumn	Rather tender ground-covering plants for moist, well-drained position in sun. Cuttings can be taken to overwinter in frame.

PLANT NAME	METHOD OF PROPAGATION	TREATMENT REQUIRED	TIME OF YEAR PROPAGATED	CULTURAL NOTES
Paronychia (Nailwort/Whitlow-wort)	SEED	Sow in prepared bed in sun outdoors. Plant out in autumn.	March/April	Carpeting plants similar to thyme for hot dry positions. Need overhead protection in winter in damp areas. Make good carpet for dwarf bulbs.
	TIP CUTTINGS	Root in cold frame or greenhouse and overwinter there. Plant out in spring 25–30 cm apart.	Late summer	
Parrya	SEED	Sow in pots in cold frame and grow on there until established.	As soon as ripe	Low-growing tufted plants related to wallflowers. Require constant propagation as they are short-lived. Some require overhead winter protection.
	STEM CUTTINGS	Root in cold frame and overwinter there. Plant out in spring.	During summer	
Patrinia	SEED (perennial often grown as biennials)	Sow in a prepared bed outdoors, thin and plant in final positions in autumn.	April	Herbaceous perennials, some suited to wild gardens but *P. intermedia sibirica* and *triloba* are suitable for the rock garden where they flower in summer.
	DIVISION	Replant in suitable positions.	Autumn or spring	
Petrocallis (P. pyrenaica = Rock Beauty)	HEEL CUTTINGS	Cuttings are made from small branches pulled from the plant. Root in cold frame and grow on there.	August	A minute cushion plant spreading about 30 cm. Grow in alpine house or moraine conditions in sun.
Phyteuma (Horned Rampion)	SEED (sow in gritty limestone compost)	Place alpine species out to stratify bringing into heat in March. Grow on in frames.	As soon as ripe or December/January	Plants for the rock garden or front of the border. Alpine species require scree conditions and/or alpine house protection.
	DIVISION	Replant according to species.	Autumn or spring	
Polygala (Milkwort; P. chamaebuxus = Bastard Box)	SOFTWOOD CUTTINGS (hardy species)	Root in cold frame or propagator in sand/peat compost and overwinter in frames. *P. chamaebuxus* spreads by stolons which can be removed and treated as Irishman's cuttings.	During summer and early autumn	Genus contains a number of evergreen prostrate shrubs for almost anywhere on the rock garden that is not too dry.
Pratia	SEED	Sow in propagator or greenhouse at 18–24°C. Grow on under glass, hardening off before planting out in June.	In spring	Creeping mat-forming plants grown mainly for their berries. A shady position suits *P. angulata* and *P. arenaria*.
	SOFTWOOD CUTTINGS	Root in warm propagator or greenhouse and overwinter under glass.	In spring or autumn	
Primula (P. vulgaris, syn. P. acaulis = Primrose)	SEED (rinse seed in cold fresh water before sowing)	Sow in pots and place outdoors to stratify. Transfer to slight heat in March. Grow on in frame or prepared bed outdoors. Plant out in autumn.	As soon as ripe or December/January	A vast family of beautiful species many suitable for rock garden and alpine house. Most 'bogland' types need sun, others are happy with semi-shade in damp woodland type soil.
	DIVISION	Divide established clumps and replant according to species.	After flowering or in autumn	
Pterocephalus	DIVISION	Carefully divide older plants and replant where they will flower.	In spring	*P. parnassi* is a low-growing tufted plant for well-drained rock garden or dry wall.
Pulsatilla (Pasque Flower)	SEED	Sow in prepared bed outdoors or in pots placed out to stratify and brought into slight heat in March. Plant out in autumn 20–23 cm apart.	As soon as ripe or in early spring	Similar in appearance to anemones. Require well-drained sunny position with much lime. May take some years to reach flowering size.

PLANT NAME	METHOD OF PROPAGATION	TREATMENT REQUIRED	TIME OF YEAR PROPAGATED	CULTURAL NOTES
Pyrola *(genus contains Wintergreens)*	SEED	Sow thinly where they will flower outdoors in prepared bed.	March/April	Plants for semi-shaded positions in peaty woodland type soil. Resents disturbance but can be divided with care in spring.
Ramonda	SEED	Sow in slight warmth and grow on in frame planting out in autumn.	February to April or August/September	Rosette-forming plants requiring a leaf-mould type soil in a north-facing position in a bank or wall. Ideal alpine house plant.
	LEAF CUTTINGS	Use mature undamaged leaves with stalk. Root in sand/peat compost in shady spot in greenhouse.	During summer	
Raoulia	DIVISION	Divide and replant in sunny well-drained position.	Late summer or autumn	Foliage plants forming low-growing mats. Ideal for troughs and sinks.
Rhodohypoxis	DIVISION (rhizomatous corms)	Divide established plants and replant 10–15 cm apart in well-drained lime-free soil or compost.	September	Hairy-leaved plants requiring overhead winter protection or alpine house culture.
Sagina *(Pearlwort)*	SEED	Sow in cold frame and grow on in small pots.	In spring	Low-growing cushion or trailing plants ideal for sinks, troughs and crevices between paving in sunny positions.
	DIVISION	Divide and replant older plants.	In spring	
Sanguinaria	DIVISION	Divide and replant in open ground 12–15 cm apart.	In August	A herbaceous low-growing plant for a leafy/sand soil in semi-shade.
Saxifraga *(Rockfoil; S. umbrosa and sub-species = London Pride; S. stolonifera, syp. S. sarmentosa = Mother of Thousands, grown as a house plant)*	DIVISION	Many types can be divided replanting according to type.	After flowering	A large genus of some 300 species divided into several sections. Grow Kabshias and Englerias in alpine house or outdoors on north-facing slope topdressed with limestone chippings. Aizoon or encrusting saxifrages require well-drained spot in sun. Mossy saxifrages such as *S. hypnoides*, a native species, require a moist semi-shaded position.
	BASAL CUTTINGS (usually side rosettes. All but mossy types)	Root under glass in propagator or cold frame depending on time of the year. Grow on in frames until established.	After flowering	
	SEED (where available)	Sow in pots and place out to stratify. Place in slight heat in March. Grow on in frames until large enough to plant out.	December/January	
Schizocodon	SEED	Sow in lime-free peat-based compost in cold frame and grow on there planting out in autumn.	As soon as ripe or in spring	Evergreen mat-forming plants requiring cool woodland conditions of semi-shade in a lime-free leafy soil.
Scutellaria *(Scull-caps. Hardy species)*	DIVISION	Replant in growing positions 30–45 cm apart.	March	Grow well in any open situation in well-drained soil. Flower in late summer and autumn.
	SEED	Sow in cold frame or greenhouse and grow on there. Plant out in September.	March/April	
Sempervivum *(House Leek)*	DIVISION (offsets)	Detach rosettes and grow on in pots until root system is well established.	Spring through to autumn	Ideal succulent plants for troughs and walls where they will survive with little attention. All are evergreen and spread their rosettes from 13 to 25 cm depending on species. Seedlings are variable and selected types are best propagated vegetatively.
	SEED	Sow in seed compost mixed with old mortar rubble placed out to stratify. Transfer to slight heat in March. Grow on in small pots in frame.	December/January	
Shortia	DIVISION	Divide carefully to avoid damage to roots. Replant 30–38 cm apart.	May/June	Evergreen sub-shrubs requiring a moist soil with peat or leaf-mould in a semi-shaded spot. An ideal alpine house plant. Can also be raised from seed when available.
	BASAL CUTTINGS	Root in peat/sand compost in shady cold frame and overwinter there.	June/July	

PLANT NAME	METHOD OF PROPAGATION	TREATMENT REQUIRED	TIME OF YEAR PROPAGATED	CULTURAL NOTES
Silene (Catchfly/Campion. S. acaulis = Moss Campion)	SEED	Sow annuals where they will flower and thin to 15 cm. Sow perennials in the cold frame and plant out in September.	March/April March	Many species are mat-forming plants for a sunny position. Overwinter cuttings in frames or under cloches. Perennials can also be divided replanting 30–38 cm apart.
	HEEL CUTTINGS	Root in cold frame or shady spot outdoors.	After flowering	
Soldanella (Moonwort)	SEED	Sow in sand/peat compost in cold frame and overwinter there.	As soon as ripe or in spring	Rhizomatous mat-forming herbaceous plants for a cool moist position. Require overhead protection from wet in winter. Spread from 15 to 30 cm depending on species.
	BASAL CUTTINGS	Root in cold frame. Water sparingly and plant out in September the following year.	May/June	
	DIVISION	Divide carefully and replant outdoors or pot and overwinter in cold frame.	June	
Swertia (S. perennis = Marsh Felwort)	SEED	Sow in well-drained peat-based compost in cold frame. Plant out in final positions in June with as little disturbance as possible.	March/April	Plants for moist rockeries and damp shady places which never dry. S. perennis is a low-growing bog plant.
Symphyandra	SEED	Sow in a warm greenhouse or propagator and grow on in frames. Plant out in September.	March	Plants of the Campanula family flowering in late summer. Require a rich sandy loam in sun. Rockery or border plants.
	DIVISION	Replant in growing positions.	Spring or autumn	
Synthyris	SEED	Sow in cold greenhouse or frame. Grow on in shady prepared bed outdoors and plant out in autumn.	March/April	Herbaceous perennials for semi-shaded woodland type soil. Woolly-leaved types need overhead winter protection or alpine house.
	DIVISION (green-leaved types. Rhizomatous roots)	Divide carefully and replant in suitable positions a few cm apart or in rock crevices.	Early autumn or after flowering	
Tanakaea (Japanese Foam Flower)	LAYERING (natural from runners)	Separate rooted runners and replant 15–18 cm apart where they will flower.	March	An evergreen woodland plant for a peaty soil in semi-shade. Male and female plants, only male producing runners.
Teucrium (Germander) (T. marum = Cat Thyme)	SOFTWOOD CUTTINGS (shrubby types)	Root in propagator or greenhouse. Pot on and grow in plunge beds until following spring.	Late spring or early summer	Some species are aromatic-forming sub-shrubs ideal for sheltered rock garden or alpine house.
	SEED	Sow in prepared beds outdoors. Thin or transplant 30 cm apart.	March/April	
Thlaspi (T. latifolium, syn. Pachyphragma macrophyllum)	SEED	Sow in well-drained compost and place out to stratify. Place in slight heat in March and grow on in frame.	As soon as ripe or December/January	Short-lived or low-growing plants for moraine or scree conditions in full sun. Propagate annually for best results.
	SOFTWOOD CUTTINGS (perennials)	Root severed rosettes in cold frame and grow on there.	April/May	
Townsendia	SEED	Sow in cold frame or prepared bed outdoors. Plant out autumn or spring.	As soon as ripe or in spring	Mat-forming plants with aster-like flowers. Requires a sunny well-drained spot or alpine house.
Trachelium (T. caeruleum = Blue Throatwort)	SEED	Sow in greenhouse or propagator at 13–18°C. Grow on in cold frame and plant out May or June.	In spring	Tender herbaceous perennials for alpine house, sunny rock gardens or dry walls. Protect from severe frosts.
	SOFTWOOD CUTTINGS	Root in cold frame and overwinter there.	April or August	

PLANT NAME	METHOD OF PROPAGATION	TREATMENT REQUIRED	TIME OF YEAR PROPAGATED	CULTURAL NOTES
Trifolium *(Clover/Shamrock)*	SEED	Sow in open ground where they will flower or in prepared bed outdoors for growing on.	August	A number of species including *T. alpinum, badium* and *uniflorum* make good rock garden plants in ordinary soil in sun.
	DIVISION (perennials)	Replant in open ground allowing for 46–60 cm spread.	March	
Tunica	SEED	Sow in cold frame and grow on in pots or prepared beds outdoors until autumn.	March	Tufted herbaceous perennials with wiry stems for well-drained position in sun or in dry walls. *T. saxifraga* will self-seed.
	DIVISION	Older plants can be divided replanting where they are to grow.	March	
Vancouveria *(V. planipetala) =* *Redwood Ivy)*	DIVISION (rhizomatous)	Divide carefully and replant where they will grow.	In spring	Groundcover plants for woodland conditions. Semi-shade in leafy soil. *V. planipetala* and *V. chrysantha* are evergreen, *V. hexandra* dies back in autumn.
Wahlenbergia	SEED	Treat annuals as half-hardy sowing in heated greenhouse or propagator. Harden off before planting out. Sow perennials in cold frame and grow on there.	February/March	A large genus closely related to and intermingled with *Edraianthus*. Plants for alpine house or scree conditions though *W. hederacea*, a native species, is a bog plant for cool moist places, being best sown where it will grow.
	BASAL CUTTINGS	Cut shoots below ground level and root in sandy compost in frame.	April/May	
Waldsteinia *(W. fragarioides =* *Barren Strawberry)*	DIVISION	Replant in growing positions.	Autumn or spring	Evergreen ground-cover plants growing well in sun or semi-shade over banks or walls.
	SEED	Sow in cold frame or prepared bed outdoors and grow on until planting out in autumn or spring.	April to July	
Zauschneria *(Z. californica =* *Californian Fuchsia)*	TIP CUTTINGS	Root in propagator or cold frame and overwinter there. Plant out in spring 30 cm or so apart.	Late summer or autumn	Sub-shrubby perennials for a light soil and sunny position where they will flower in autumn. Require frost protection in colder areas.

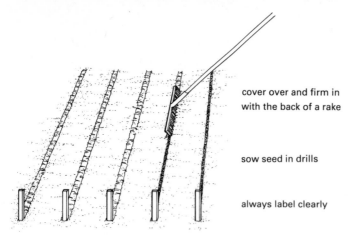

cover over and firm in
with the back of a rake

sow seed in drills

always label clearly

Biennials for bedding are best sown in nursery beds
in summer and thinned or transplanted when 5–8 cm
tall, to be put in their final positions in autumn

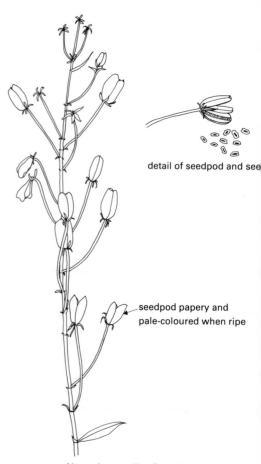

detail of seedpod and see

seedpod papery and
pale-coloured when ripe

Nemesia – seeding flowerhead.
Seed can be saved from
many mixed types of annuals

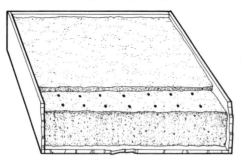

Sow pelleted seed at the same distance
apart as conventional seedlings would
be pricked out (i.e. 42 to a box)

sow seed in lines to help identify
seedlings as they emerge

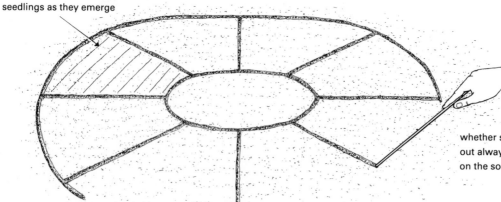

whether sowing direct or planting
out always mark out any design first
on the soil surface with a cane or stick

9 Bedding Plants

The term 'bedding plants' is used to describe all those plants used to give a temporary display of flowers in the garden—usually for the summer months, although spring bedding is also popular. Climbers have also been included, even though they seldom form part of a formal bedding scheme. This section includes all hardy annuals (HA), half-hardy annuals (HHA) and many of the hardy biennials (HB) which are used either in bedding schemes or among other plants in the mixed border. Perennials best treated as annuals, such as antirrhinums, have been included, but bulbs used for bedding, such as hyacinths, have not been included here. The abbreviations HA, HHA and HB indicate the best *treatment* and do not necessarily reflect the plant's true longevity or hardiness.

Flower beds to receive these plantings/sowings should be prepared well in advance, using rotted manure or a scattering of bonemeal to maintain fertility, rather than quicker-acting fertilisers which may make for larger plants but cause a reduction in the number of flowers produced.

For sowing direct the soil needs to be worked up as finely as possible and made level before sowing. It is a good plan to mark out exactly where an item is to be sown before starting, lines being drawn in the soil with a short cane. In this way quite complicated floral designs can be undertaken with confidence. Most bedding plants flourish best in light soils in full sun. For container planting use a good compost, John Innes Potting Composts Nos. 2 or 3, or their equivalent, being ideal. Where plants are raised under glass pay particular attention to hardening them off well before planting them out when the danger from frost has passed.

PLANT NAME	METHOD OF PROPAGATION	TREATMENT REQUIRED	TIME OF YEAR PROPAGATED	CULTURAL NOTES
Acroclinium *HA (syn. Helipterum)*	SEED	Sow in flowering position and thin to 15 cm apart.	April/early May	Height 38–45 cm. Grow in warm sunny spot. Flowers can be dried as 'everlastings'.
Ageratum *HHA (Floss Flower)*	SEED (minute)	Sow in slight heat and plant out in late May/early June 15–20 cm apart.	March/April	Height 15–20 cm. Grow in ordinary soil in sun. Bedding, edging or container plants.
Agrostemma *HA (Corn Cockle)*	SEED	Sow in flowering position and thin to 23–30 cm.	March/April or August/September	Height 75–90 cm. Flowers are good for cutting. Sunny position.
Alonsoa *HHA (Mask Flower)*	SEED	Sow in slight heat and plant out late May/early June 30–45 cm apart.	March or August for flowering under glass	Height 30–60 cm. Half-hardy herbaceous perennials grown as HHA. Can be grown in pots in heated greenhouse.
Alyssum *HA (Lobularia)*	SEED	Sow in slight heat and plant out late May/early June or in flowering position.	March/April April/May	Height 8–15 cm. Edging plants for sunny positions. Plant or thin 15–25 cm apart.
Amaranthus *HHA/HA (A. caudatus = Love-lies-Bleeding. A. hypochondriacus = Prince's Feather)*	SEED	Sow in slight heat and prick out singly into small pots. Plant out late May/early June or sow where they will flower.	March/April May	Height 60–75 cm. Specimen or dot plants for bedding schemes or groups in border. Rich soil in sunny position. Space out 23–30 cm apart.
Ammobium *HHA*	SEED	Sow in slight heat and plant out 23 cm apart in late May/early June.	February/March	Height 45–60 cm. Cut flowers can be treated as 'everlasting'. Can also be sown in September and overwintered under frost-free glass.
Antirrhinum *HHA (A. majus = Snapdragon; A. maurandioides is a perennial climber)*	SEED	Sow in a temperature of 15–18°C and plant out late May/early June 30–45 cm apart.	February/March	Height 15–90 cm. Bedding or border plants in sun. Can be sown in August to flower in cold greenhouse in spring.
Arctotis *HHA (African Daisy)*	SEED	Sow in slight heat and plant out late May/early June 30 cm apart.	March/April	Height 30–45 cm. Ideal for border or cutting in dry sunny positions.
Atriplex *HA (A. hortensis rubra = Red Mountain Spinach)*	SEED	Sow in flowering positions and thin to 45 cm.	April/May	Height 1.2–1.5 m. Foliage plant for cutting or temporary hedge or screen. Leaves can be eaten as spinach.
Bartonia *HA (B. aurea, syn. Mentzelia lindleyi)*	SEED	Sow in flowering position and thin to 45 cm.	March to May	Height 45 cm. Border plants for rich soil in sun.
Begonia *(fibrous-rooted) (B. semperflorens is an HHP treated as an HHA)*	SEED (surface sow)	Sow in a temperature of 15–18°C and grow on in frost-free greenhouse. Plant out late May/early June 15–20 cm apart.	January/February	Height 15–40 cm. Ideal for bedding in peaty soil in sun or semi-shade. Also make good pot plants. Can be lifted and overwintered under frost-free glass.
Brachycombe *HHA (Swan River Daisy)*	SEED	Sow in slight heat and plant out late May/early June 15–23 cm apart.	March/April	Height 30 cm. Grow in dry sunny positions.
Calandrinia *HHA (Rock Purslane)*	SEED	Sow in flowering positions and thin to 15–18 cm.	Late April/May	*C. umbellata* is ideal for rockery or edging in sunny position. Taller types front of the border. Dislike transplanting.

PLANT NAME	METHOD OF PROPAGATION	TREATMENT REQUIRED	TIME OF YEAR PROPAGATED	CULTURAL NOTES
Calceolaria *HHA (Slipperwort)*	SOFTWOOD CUTTINGS (*C. integrifolia* HHP) SEED	Root in cold frame and protect from frost over winter. Sow in slight heat and plant out late May/early June.	August/September February/March	Height 30 cm. Bedding or container plants for sun or semi-shade. Plant out 15–23 cm apart.
Calendula *HA (Pot Marigold)*	SEED	Sow in flowering positions and thin to 30–45 cm.	March/April or August/September	Height 30–60 cm. Grows in most situations, though prefers sun. Will self-seed.
Callistephus *HHA (Chinese Aster)*	SEED	Sow in slight heat and plant out 23–45 cm apart late May/early June, or sow where they will flower.	March/April May	Height 15–76 cm depending on variety. Plants for bedding, border or cutting.
Campanula *(C. medium = Canterbury Bells. HB)*	SEED	Sow in prepared beds outdoors. Thin to 15 cm and plant out in final positions 30 cm apart in autumn.	May/June	Height 46–76 cm. Grow in a sunny position and protect overwinter in colder areas.
Cardiospermum *HA (Balloon Vine)*	SEED	Sow where they will flower placing seeds individually 30–45 cm apart.	April/May	Height 3 m. Tendril climbing plant for covering trellises etc.
Celosia *HHA (Cockscomb)*	SEED	Sow in temperature of 18–21°C and prick out into individual pots to grow on.	March/April	Height 30–60 cm. Grow in rich soil in sun. Plant out 23 cm apart in late May/early June. Also good in pots under glass.
Centaurea *HA (C. cyanus = Cornflower; C. moshata = Sweet Sultan)*	SEED	Sow in flowering position and thin to 30 cm.	March/April	Height 30–75 cm. Plants for mixed borders or cutting in sun or sémi-shade.
Cheiranthus *HB (C. cheiri = Wallflower)*	SEED	Sow in prepared bed outdoors and thin to 15 cm. Plant in final positions 20–30 cm apart in autumn.	May to July	Height 23–45 cm. Scented plants for spring bedding especially with tulips.
Chrysanthemum *HA species*	SEED	Sow in a sunny spot where they will flower.	April	Height 15–75 cm depending on type. Thin to 23–38 cm.
Cladanthus *HA*	SEED	Sow in flowering positions and thin to 20 cm.	March/April	Height 76–90 cm. Daisy-like flowers for a warm sunny spot.
Clarkia *HA*	SEED	Sow in flowering positions and thin to 23–30 cm.	March/April or August/September	Height 38–60 cm. Plant in well-drained spot in full sun. Good for cutting.
Cleome *HHA (Spider Flower)*	SEED	Sow in slight heat and plant out 45 cm apart in late May/early June.	March/April	Height 90 cm. Border plants for rich moist soil in sun.
Cobaea *HHA (Cup and Saucer Vine; C. scandens = Cathedral Bells)*	SEED (sow on edge)	Sow in a temperature of 21–24°C and prick out singly in pots. Plant out in late May/early June 46 cm apart.	February/March	Climbing plants for a south or west-facing position in favourable districts. HHP grown as HHA. Can grow to cover an area of 2.5 m in a season. Best seen as a greenhouse climber.
Collinsia *HA*	SEED	Sow where they will flower and thin to 10–15 cm.	March to May or September	Height 23 cm. Front border plants for sun or slight shade.
Cosmos *HHA*	SEED	Sow in slight heat and plant out late May/early June 30–45 cm apart. Sow where it will flower.	February/March. May	Height 60–90 cm. Plants for sun or semi-shade in border or for cutting.
Crepis *HA*	SEED	Sow in flowering positions and thin to 15–18 cm.	March/April	Height 30 cm. *C. rubra* grows well in rock garden or front of the border.

PLANT NAME	METHOD OF PROPAGATION	TREATMENT REQUIRED	TIME OF YEAR PROPAGATED	CULTURAL NOTES
Curcubita *HHA (Gourds)*	SEED	Sow singly in small pots in slight heat and plant out late May/early June. Sow where they will grow.	March/April Late May	Tall climbing plants grown for non-edible decorative fruits which can be varnished when dried.
Dahlia *HHA (bedding types)*	SEED	Sow in slight heat and plant out late May/early June 30–60 cm apart.	March/April	Height 30–60 cm. Bedding or border plants for sun or semi-shade.
Datura *HHA types (D. stramonium, Thorn Apple, is a native species)*	SEED	Sow in slight heat and prick out singly in small pots. Plant out 45–60 cm apart in late May/early June.	March/April	Height 75–90 cm. Showy border plants. *D. stramonium* is an HA and can be sown where it will flower in late April/May, but bear in mind that this is a poisonous plant.
Delphinium *HA (D. ajacis and D. consolida = Larkspur)*	SEED	Sow in flowering positions and thin to 30 cm.	March/April or August/September	Height 45–90 cm. Border plants or for cutting.
Didiscus *HHA (Blue Lace Flower)*	SEED	Sow in a temperature of 15–18°C and plant out 25–30 cm apart late May/early June.	March/April	Height 40–45 cm. Plant in ordinary soil in sunny sheltered spot or in 13 cm pots in greenhouse.
Dimorphotheca *HHA/HA (Star-of-the Veldt)*	SEED	Sow in slight heat and plant out late May/early June or sow where they will flower and thin to 30 cm.	March/April May	Height 15–45 cm. Bedding or cut flower. Flowers only open in sunshine.
Eccremocarpus *HHA (E. scaber = Chilean Glory Flower)*	SEED	Sow in slight heat and prick out singly into small pots. Plant out late May/early June.	March	HHP climbing plants raised as HHA. For warm position in favourable areas in good soil. An established plant grows to 3 m.
Echium *HA/HB (Vipers' Bugloss)*	SEED	Sow in flowering positions and thin to 23–38 cm.	March/April or August/September	Height 30 cm. Border or bedding in sun or semi-shade.
Emilia *HHA/HA (E. flammea, syn. Cacalia coccinea = Tassel Flower)*	SEED	Sow in slight heat and plant out in May or sow where they flower, thinning to 15–20 cm.	March/April Late April/May	Height 30 cm. Good in mixed borders.
Eschscholtzia *HA (E. californica = Californian Poppy)*	SEED	Sow in flowering positions and thin to 30 cm.	March/April or August/September in mild areas	Height 23–30 cm. Border or rock garden plants for sunny spot.
Euphorbia *HHA/HA (E. marginata = Snow-on-the-Mountains)*	SEED	Sow in slight heat and plant out late May or sow where they will grow. Thin to 30 cm.	March/April April/May	Height 45–75 cm. Foliage plants for mixed borders or cutting for arrangements.
Felicia *HHA (Kingfisher or Blue Daisy)*	SEED	Sow in slight heat and plant out late May/early June 23–30 cm apart.	March/April	Height 10–30 cm. Border or bedding in sun or slight shade. Shrubby types are greenhouse perennials.
Gaillardia *HHA/HA (Blanket Flower)*	SEED	Sow in slight heat and plant out late May/early June or sow where they will flower and thin to 30 cm apart.	March/April April/May	Height 45 cm. Plants for cutting or space filling in the mixed border in sunny positions.
Gilia *HA (Bird's Eyes)*	SEED	Sow in flowering positions and thin to 30 cm.	March/April	Height 30–60 cm. Rock gardens or front of the border plants. *G. coronopifolia* is treated as an HHA.
Godetia *HA*	SEED	Sow in flowering positions and thin to 23–38 cm.	March/April or August/September	Height 23–60 cm. Grow tall types for cutting and dwarf for bedding.

PLANT NAME	METHOD OF PROPAGATION	TREATMENT REQUIRED	TIME OF YEAR PROPAGATED	CULTURAL NOTES
Gomphrena *HHA (G. globosa = Globe Amaranth)*	SEED	Sow in slight heat and plant out 15–23 cm apart in late May/early June.	March/April	Height 15–30 cm. Edging or border plants or in 13 cm pot in greenhouse. Flowers can be dried as 'everlasting'.
Grass *HA (ornamental types. Briza maxima = Quaking Grass)*	SEED	Sow where they are to grow and thin to 15–30 cm. Grow in sun or slight shade.	March to May	Height 30–60 cm. Mainly grown for cutting either fresh or for dried winter floral arrangements.
Gypsophila *HA (Chalk Plant/Baby's Breath)*	SEED	Sow in flowering positions and thin to 23–30 cm.	March onwards for succession or in autumn	Height 30 cm. Prefers chalk or limestone soils. Borders or for cutting.
Helianthus *HA (Sunflower)*	SEED	Sow singly in small pots in slight heat and plant out in May or sow 60–90 cm apart outdoors where they will grow.	February/March

April | Height 0.5–2.5 m. Tall plants for the back of borders in rich soil in sun or slight shade. |
| **Heliophila** *HHA/HA* | SEED | Sow in slight heat and plant out 30 cm apart in late May/early June or sow where it will flower. | March/April

Late May | Height 45 cm. Border plants or pot plants in the greenhouse. |
| **Heliotropium** *HHP grown as an HHA (Cherry Pie/Heliotrope)* | SEED

SOFTWOOD CUTTINGS | Sow in a temperature of 15–18°C. Grow on in warmth and plant out in June.

Root in propagator, pot singly and grow on as for seeds. | February/March

In spring | Height 45–60 cm. Scented flowers for borders or plant containers in warm spots. Can be potted and overwintered in warm greenhouse or grown there permanently. |
| **Hibiscus** *HHA/HA (H. trionum = Bladder Katmia/Flower-of-an-Hour)* | SEED | Sow in slight heat and plant out end of May 30–38 cm apart, or sow where it will flower. | March/April

Late May | Height 45 cm. Grow in sunny spot in poor soils. Other HHA types need similar treatment. |
Humulus *HHA/HA (H. japonicus = Japanese Hop)*	SEED	Sow in slight heat and prick out singly into small pots. Plant out late May. Can also be sown where it will grow in May.	March	Annual climber ideal for arches etc.
Hunnemannia *HHA (Mexican Tulip Poppy)*	SEED	Sow in slight heat and plant out 30 cm apart in late May/early June.	March/April	Height 60 cm. Useful for cutting or growing in borders.
Iberis *HA (Candytuft)*	SEED	Sow in flowering positions and thin to 15–23 cm.	March to May or August/September	Height 23–30 cm. Ideal for rockeries, borders or bedding.
Impatiens *HHA (Busy Lizzie; I. balsamina = Balsam)*	SEED (sow shallowly)	Sow in a temperature of 21–24°C in shade. Grow on in warmth and harden off gradually before planting out 23–30 cm apart in late May/early June.	February/March	Height 15–60 cm, according to species. Modern strains of Busy Lizzies make ideal bedding or container plants. Balsam is very different in habit and appearance. Grow well in shade or semi-shade.
Inopsidium *HA*	SEED	Sow in flowering positions and thin or transplant to 10–12 cm apart.	March to May	Low-growing plants suitable to rockeries or crevices in paving in sun or shade.
Ipomoea *HHA (Morning Glory)*	SEED (hard-soak 24 hours in tepid water before sowing)	Sow singly in small pots in a temperature of 15–18°C and plant out early June.	March/April	Climbing plants for warm sunny positions outdoors or in the greenhouse or conservatory.
Kochia *HHA (Burning Bush)*	SEED	Sow in slight heat and plant out 45 cm apart in late May/early June.	March/April	Height 30 cm. Foliage dot plant for bedding schemes.

PLANT NAME	METHOD OF PROPAGATION	TREATMENT REQUIRED	TIME OF YEAR PROPAGATED	CULTURAL NOTES
Lagenaria HHA (L. vulgaris = Bottle Gourd)	SEED (soak 48 hours in tepid water before sowing)	Sow singly in small pots in a temperature of 24–27°C and plant out in late May/early June.	March/April	Vigorous climbers yielding inedible ornamental gourds for many craft purposes. Grow in rich soil in sunny position.
Layia HA (Tidy Tips)	SEED	Sow in flowering positions and thin to 20–23 cm.	April/May	Height 30 cm. Flowers are good for cutting.
Leptosiphon HA (syn. Gilia lutea)	SEED	Sow in flowering positions and thin to 10–12 cm.	April/May	Height 10–15 cm. Edging, rockery and carpeting plants.
Limnanthes HA (Poached Egg Plant)	SEED	Sow in flowering positions and thin to 15 cm.	March/April or August/September	Height 15 cm. Edging plants for sunny positions.
Linaria HA (Toadflax)	SEED	Sow in flowering positions and thin to 15–23 cm.	March to May	Height 15–30 cm. Use in borders or rock gardens in sun.
Linum HA (L. grandiflorum = Flax)	SEED (sow at intervals for succession)	Sow in flowering positions and thin to 15–23 cm.	March to May	Height 30–45 cm. Grow in sun or semi-shade.
Lobelia HHA	SEED (seedlings small. Prick out in *small* clusters of a few seedlings)	Sow in a temperature of 18–21°C and plant out late May/early June 15–23 cm apart.	February/March	Height 10–15 cm. Edging plants or plant containers in sun. Associate well with white *Alyssum.*
Lonas HA	SEED	Sow in flowering positions and thin to 30 cm apart.	March/April	Height 38 cm. Flowers can be dried as 'everlastings'.
Lunaria HB (Honesty)	SEED	Sow in shady prepared bed outdoors and move to final positions 25–30 cm apart when seedlings are 5–7 cm high.	May/June	Height 75 cm. Grown for decorative disc-like seed capsules and as border plants. Cut stems in late summer for winter arrangements.
Malcomia HA (Virginian Stock)	SEED	Sow where they will flower. No thinning required if sown thinly.	March to June	Height 15 cm. Easily grown plants ideal for crevices in paving and odd corners.
Malope HA	SEED	Sow thinly in flowering positions and thin to 45 cm.	March/April	Height 60–90 cm. Good border or cut flower plants for ordinary soil in sun.
Matricaria HHA/HA (M. eximia, syn. Chrysanthemum parthenium = Feverfew)	SEED	Sow in greenhouse at 15–20°C. Plant out late May after hardening off. Can also be sown in flowering positions and thinned to 23 cm.	February/March April	Height 30 cm. Bedding or front border plants for light soils in sun.
Matthiola HA/HHA/HB (M. bicornis = Night-scented Stock; M. incana = Stock)	SEED	Sow Night-scented Stocks where they will flower and thin to 15–23 cm. Sow ordinary summer stocks under glass and plant out in late May or early June. Sow Brompton Stocks in summer for spring flowering.	March onwards until summer February/March July	Height 30–45 cm. Scented flowers for sun or shade.
Maurandia HHP treated as HHA	SEED (winged, slow to germinate)	Sow singly in small pots in a temperature of 15–21°C and plant out 45 cm apart in late May/early June.	March/April	Climbing plants which can be overwintered in frost-free greenhouse if desired. Grow in warm position against walls etc or in greenhouse.
Mesembryan-themum HHA (M. criniflorum = Livingstone Daisy)	SEED	Sow in slight heat and plant out 23 cm apart in late May/early June.	February/March	Low-growing plants ideal for edging or banks in dry sunny positions. Flowers only open in sun.

PLANT NAME	METHOD OF PROPAGATION	TREATMENT REQUIRED	TIME OF YEAR PROPAGATED	CULTURAL NOTES
Mirabilis HHA (M. jalapa = Marvel-of-Peru/Four o'clock)	SEED	Sow in slight heat and plant out 30–38 cm apart in late May/early June.	March/April	Height 60–90 cm. Bushy plants for borders. Develop tubers which can be lifted and stored like dahlias.
Moluccella HHA (Bells-of-Ireland/Shell Flower)	SEED (germination slow)	Sow in a temperature of 21–24°C and plant out 23–38 cm apart in early June or sow where they will flower.	April Late May	Height 45 cm. Grown for floral decoration as cut flower or dried as 'everlasting'.
Nemesia HHA	SEED	Sow in slight heat and plant out 15–23 cm apart in late May/early June.	March/April	Height 23–30 cm. Front of the border or bedding plants. Prefers moist soil in sun.
Nemophila HA (Baby Blue Eyes)	SEED	Sow where they will flower and thin to 23 cm.	March to May or August/September	Trailing plants for sunny positions. Make good pot plant in cold greenhouse.
Nicandra HA (N. physaloides = Shoo-Fly-Plant)	SEED	Sow outdoors where it will flower and thin or transplant to 60 cm apart.	April/May	Height 90 cm. Border plants. Stems in fruit can be dried for winter decoration.
Nicotiana HHA (Flowering Tobacco)	SEED	Sow in slight heat and plant out 45–60 cm apart in late May/early June.	March/April	Height 30–75 cm. Scented plants for border or bedding in sun or semi-shade.
Nigella HA (Love-in-a-Mist)	SEED	Sow where they will flower and thin to 23 cm apart.	March/April or August/September	Height 23–45 cm. Bedding or border plants for sun or semi-shade.
Nolana HHA (Chilean Bell Flower)	SEED	Sow under glass or in flowering positions outdoors and thin/plant out 30–38 cm apart.	March/April, late April/May	Low-growing trailing plants for sunny, dry positions in rock garden or border.
Ornamental Cabbage etc HHA/HA	SEED	Sow under glass or where they will grow and thin or plant out 30–38 cm apart.	March/April May	Height 30–60 cm. Colourful and unusual bedding plants or in border groups for effect.
Perilla HHA	SEED	Sow in a temperature of 15–18°C and plant out 30 cm apart late May/early June	February/March	Height 45–60 cm. Foliage plants for specimen or spot position in summer bedding schemes.
Petunia HHA	SEED	Sow in slight heat and plant out 23–38 cm apart in late May/early June.	March/April	Height 23–45 cm. Ideal for bedding or containers in sun or slight shade.
Phacelia HA	SEED	Sow in flowering positions and thin to 15 cm.	April/May or August/September	Height 23–45 cm. Border, edging or rock garden plants for sunny positions.
Phlox HHA (P. drummondii)	SEED	Sow in slight heat and plant out 23 cm apart in late May/early June or sow where they will flower.	March/April May	Height 15–30 cm. Border, bedding or edging plants for sun or semi-shade.
Platystemon HA (Cream Cups)	SEED	Sow in flowering positions and thin to 18–20 cm.	March/April or August/September	Height 30 cm. Border plants for well-drained soil in sun.
Portulaca HHA (Rose Moss)	SEED	Sow in the greenhouse at 15–20°C and plant out after hardening off. Can also be sown in open ground in favourable districts.	March May	Carpeting plants for rock garden or paving. Grow *in situ* as it resents transplanting. Flowers only open in sunshine.
Quamoclit HHA (Q. coccinea = Star Glory)	SEED (soak 24 hours in tepid water before sowing)	Sow singly in small pots at 18–21°C and plant out in early June.	March/April	Climbers for a warm sunny spot outdoors or in the greenhouse. *Q. coccinea* grows some 2.5 m tall.

PLANT NAME	METHOD OF PROPAGATION	TREATMENT REQUIRED	TIME OF YEAR PROPAGATED	CULTURAL NOTES
Reseda HA (Mignonette)	SEED	Sow in flowering positions and thin to 23–30 cm.	April onwards	Height 30 cm. Scented and ideal for cutting. Requires a well-drained soil in sun with plenty of lime.
Rhodanthe HA (Swan River Everlasting, syn. Helipetrum)	SEED	Sow in flowering positions and thin to 15 cm.	April/early May	Height 30–38 cm. Flowers can be dried as 'everlastings'.
Ricinus HHA (Castor Oil Plant/Castor Bean)	SEED (poisonous)	Sow singly in small pots in slight heat and plant out 75 cm apart in late May/early June.	February/March	Height up to 2.5 m. Foliage plants for the back of borders or as specimen plants.
Salpiglossis HHA (Painted Tongue)	SEED	Sow in slight heat and plant out 30 cm apart in late May/early June.	March to May	Height 60–90 cm. Borders or for cutting. Flowers well in pots in the greenhouse.
Salvia HHA (S. splendens = Scarlet Sage)	SEED	Sow in slight heat prick out singly and grow on at 15°C. Plant out 30 cm apart in late May/early June.	February/March	Height 23–75 cm. Bedding and container plants for a rich soil in sun. S. patens can be overwintered in heated greenhouse.
Santvitalia HA/HHA (S. procumbens = Creeping Zinnia)	SEED	Sow in slight heat and plant out 25–30 cm apaft in late May/early June.	March/April	Low-growing edging or rockery plants for moist soils in sun.
Schizanthus HHA (Butterfly Flower)	SEED	Sow in slight heat and plant out 30–38 cm apart in early June or sow where they flower.	March onwards Late May	Height 30–45 cm. Border plants, cut flower or cool greenhouse pot plant.
Silybum HA (Our Lady's Milk Thistle)	SEED	Sow in flowering positions and thin to 45–60 cm.	April/May	Height 1.2 m. Foliage plants for border or specimen spot. Remove flowers as they form.
Tagetes HHA (T. patula = French Marigold. T. erecta = African Marigold)	SEED	Sow in slight heat and plant out late May/early June or sow where they will grow. Thin French types to 23–30 cm African to 30–60 cm.	March/April Late May	Height 15–75 cm. Edging, bedding and container plants for ordinary soil in sunny positions.
Thunbergia HHA/HHP (Black-Eyed-Susan)	SEED	Sow in slight heat and plant out 30 cm apart in late May or early June.	March/April	Climbing plants to about 1.2 m for a warm spot outdoors or as a greenhouse/house plant.
Tithonia HHA (Mexican Sunflower)	SEED	Sow in slight heat and prick out singly in small pots. Plant out in early June 45 cm apart.	March	Height 90 cm. Border plants for good soil in sun.
Tropaeolum HA (T. majus = Nasturtium. T. peregrinum, syn. T. canariensis = Canary Creeper)	SEED	Sow where they are to flower placing seeds 30 cm apart.	April/May	Climbing or trailing plants for poor soils in sun or slight shade.
Ursinia HHA	SEED	Sow in slight heat and plant out 23–30 cm apart in late May/early June or sow where they will grow.	March/April Late May	Height 30 cm. Daisy-like flowers for warm sunny borders.
Venidium HHA	SEED	Sow in slight heat and plant out 25–36 cm apart in late May/early June. V. calendulaceum can be sown in situ.	April April/May	Height 45–90 cm. Border plants in warm spot or sown August/September and grown on at 10–13°C it makes a good greenhouse pot plant.

PLANT NAME	METHOD OF PROPAGATION	TREATMENT REQUIRED	TIME OF YEAR PROPAGATED	CULTURAL NOTES
Verbena *HHP grown as an HHA*	SEED	Sow in slight heat and plant out 30 cm apart in late May/early June.	March/April	Height 15–30 cm. Bedding plants for ordinary soil in full sun.
Viscaria *HA*	SEED	Sow in flowering positions and thin to 15 cm apart.	March/April or August/September	Height 15–30 cm. Grow in the border or for cutting in well-drained positions.
Xeranthemum *HA (Common Immortelle)*	SEED	Sow in flowering positions and thin to 30 cm.	March/April	Height 60 cm. Flowers can be cut and dried as 'everlastings'.
Zea *HHA (Striped Maize)*	SEED	Sow singly in small pots in slight heat and plant out 45 cm apart in late May/early June.	February/March	Height 1.2 m. Foliage plants best in border groups for effect. Can be sown in flowering positions in May if necessary.
Zinnia *HHA*	SEED	Sow in slight heat and prick out into small pots. Plant out in late May/early June 23–45 cm apart depending on type.	March/April	Height 15–60 cm. Edging, bedding and cut flower plants for growing in full sun. Can be sown where they will flower in late May if necessary.

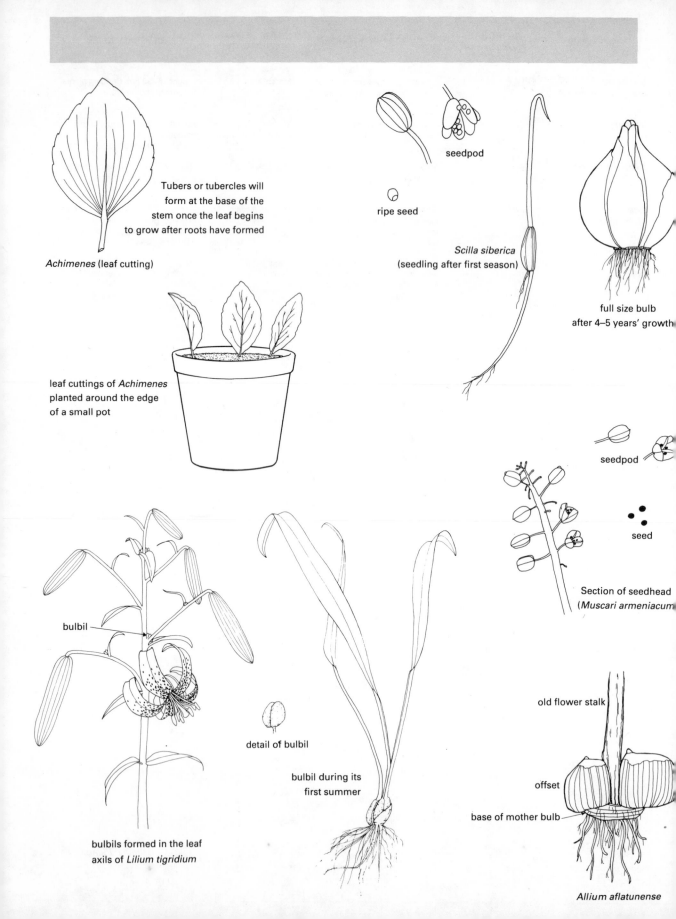

Tubers or tubercles will
form at the base of the
stem once the leaf begins
to grow after roots have formed

Achimenes (leaf cutting)

leaf cuttings of *Achimenes*
planted around the edge
of a small pot

seedpod

ripe seed

Scilla siberica
(seedling after first season)

full size bulb
after 4–5 years' growth

seedpod

seed

Section of seedhead
(*Muscari armeniacum*

bulbil

detail of bulbil

bulbil during its
first summer

bulbils formed in the leaf
axils of *Lilium tigridium*

old flower stalk

offset

base of mother bulb

Allium aflatunense

10 Bulbs

The list of bulbs in this chapter also includes many other plants which are not, strictly speaking, true bulbs, as they have rhizomatous or tuberous roots. They are, however, often treated as bulbs, both by the gardener and his suppliers, and so are included here.

By and large most bulbs grown in the garden require a really well-drained position in good light soil, and, although many of those species flowering in late autumn/ early spring are tolerant of some shade, those of a more tender nature require a position in full sun. When growing in pots or containers in greenhouse or garden room a large number of bulb species will be happy growing in a compost of equal parts (by volume) of fibrous loam, leafmould or moss peat, and well-rotted manure, with a liberal sprinkling of sand or grit.

For seed sowing, John Innes seed compost or its equivalent should prove satisfactory. Sow thinly, and in most cases leave the seedlings undisturbed for their first year of growth before transplanting. Grow smaller types on in deep boxes of sandy compost, larger hardy types in prepared beds in a sunny position outdoors. Grow the offsets on in a similar manner.

PLANT NAME	METHOD OF PROPAGATION	TREATMENT REQUIRED	TIME OF YEAR PROPAGATED	CULTURAL NOTES
Achimenes (Hot Water Plant)	SEED	Sow in a temperature of 18–24°C and grow on in heat.	May	Greenhouse plants or for the house in a sunny window. Plant 1–3 tubercles to an 8 cm pot and water with warm water.
	DIVISION (of tubercles)	Divide tubercles as soon as growth begins	February/March	
	LEAF CUTTINGS	Use mature leaves complete with stalk and root in slight heat.	June to August	
Acidanthera	DIVISION (offsets; corms)	Detach offsets when lifting and store in dry sand in warm place for the winter. Grow offsets on in boxes in spring.	October when foliage dies back	Grow outdoors in warm sunny positions or in the greenhouse. Plant out in May 8 cm deep, 15 cm apart.
Agapanthus (Blue African Lily)	DIVISION (of rhizome)	Replant divisions in chosen spot 20 cm deep, 45 cm apart, or 3 roots to large tub.	April/early May	Grow outdoors in warm positions in borders or in tubs. Protect from frost in winter. Seedlings can take up to 5 years to reach flowering size.
	SEED	Sow in a temperature of 13–16°C and grow on in greenhouse or frame.	April	
Albuca	DIVISION (offsets; bulbs)	Replant offsets in small pots and grow on in slight heat.	September	Bulbous plants best grown in a slightly heated greenhouse. Seedlings will take 4–5 years to reach flowering size.
	SEED	Sow in a temperature of 16–18°C and grow on in the greenhouse.	March/April	
Allium (Genus includes onions, chives and garlic as well as many ornamental species)	DIVISION (offsets bulbs)	Replant offsets in boxes and grow on in cold frame.	October	A large genus with many species suitable for rock garden or borders. Offsets will flower 2 years from separation, seedlings slightly more.
	SEED	Sow in slight heat and grow on as for offsets.	March/April	
Alstroemeria (Peruvian Lily)	DIVISION (of tubers)	Replant in open ground 15 cm deep 20–25 cm apart.	March/April	Best grown in border groups. Transplant carefully as roots are very brittle. Seed is slow to germinate. Grow seedlings on in cold frame for 2 years. Will flower in their third or fourth year.
	SEED (soak 24 hours in tepid water before sowing)	Sow in a gritty compost in a temperature of 18–21°C or sow freshly-gathered seed in flowering positions.	April

October | |
Amaryllis (A. belladonna = Belladonna Lily; for Amaryllis hippeastrum see Hippeastrum)	DIVISION (offsets; bulbs)	Remove offsets when repotting at 5 year intervals and grow on in heated greenhouse.	May to July	Grow outdoors in mild areas only, elsewhere in pots or tubs moved to heated glass in winter. Outside plant 15–20 cm deep, 45 cm apart. In pots plant just below compost surface.
	SEED	Sow in peaty compost in a temperature of 18°C and grow on as for offsets. Take 7 years to reach flowering size.	April to June	
Ammocharis (Malagas Lily)	DIVISION (offsets; bulbs)	Remove large offsets and grow on in cold greenhouse watering carefully until roots have developed.	May/June	Winter flowering so grow in greenhouse or conservatory 1 bulb to a 18 cm pot. Offsets take some years to reach flowering size.
Anemone (Windflower) (A. coronaria = Anemone de Caen)	SEED (Benefits from impactation before sowing)	Sow in cold frame or in drills in a prepared bed outdoors and thin to 10 cm.	As soon as ripe or March/April	Numerous types with many uses from spring flower display to cut flower. Most require lime and plenty of sun. Seed can be incubated in damp sand at 16–18°C for 10 days prior to sowing.
	DIVISION (of corms or tuber)	Older corms can be divided if required storing them in dry peat/sand until time to replant. Replant A. palmata in open ground in full sun.	When foliage has died back

September/October | |

PLANT NAME	METHOD OF PROPAGATION	TREATMENT REQUIRED	TIME OF YEAR PROPAGATED	CULTURAL NOTES
Anthericum (A. liliago = St. Bernard's Lily)	SEED	Sow in cold frame and prick out into nursery bed outdoors to grow on.	April or as soon as ripe	Rock garden or border plants growing 45–60 cm in height. Seedlings will flower 2–3 years from sowing. Divide carefully to avoid damaging the roots.
	DIVISION (tuber)	Replant in open ground 8–15 cm deep, 45 cm apart.	March/April	
Antholyza (A. paniculata, syn. Curtonus paniculatus = Aunt Liza)	DIVISION (offsets; corms)	Detach cormlets and grow on in boxes of sandy compost in greenhouse or frame.	In late summer from established groups	Half-hardy. Plant in warm position 15 cm deep and protect from frost over winter or grow in heated greenhouse. Grow 45 cm tall.
Arisaema (hardy species) (greenhouse types require 10°C winter minimum)	DIVISION (tuber)	Replant in humus-rich soil 5–8 cm deep, 30–45 cm apart.	In spring	Plants for a moist woodland setting. Associate well with ferns and hostas. Grow from 45 to 90 cm in height.
	SEED	Sow in slight heat and grow on under shaded glass until established.	March/April	
Arthropodium	DIVISION (of roots or by offsets)	Replant in open ground 23 cm apart or in pots in a slightly heated greenhouse.	March/April	Require a warm position outdoors or greenhouse culture. They make plants 30–90 cm in height. Plant or pot seedlings in following spring.
	SEED	Sow in slight heat and grow on there.	April	
Arum (A. maculatum = Lords and Ladies)	DIVISION (tuber)	Replant in humus-rich soil 38–45 cm apart.	In spring as growth begins.	Plants for woodland conditions of semi-shade. Grown for their foliage as well as their curious 'flowers'.
Asphodelus (Asphodel Lily)	DIVISION (rhizomes)	Replant 15 cm deep, 38 cm apart in deep sandy soil outdoors.	March/April	Border plants growing 90 cm or more in height. Divide carefully and only at 5-year intervals as it resents disturbance.
Babiana (Baboon Flower)	DIVISION (offsets; corms)	Remove cormlets on lifting and store in dry sand over winter. Grow on in boxes or frames the next spring.	In summer when foliage dies down	Only hardy in mildest areas, elsewhere grow 6 corms to a 13 cm pan in greenhouse. Outdoors plant 8 cm deep, 15–20 cm apart. Transplant seedlings yearly until flowering size.
	SEED	Sow in slight heat and grow on there.	As soon as ripe	
Begonia, Tuberous-rooted	SEED (dust-like)	Sow in a temperature of 18°C covering box with glass and paper until germination occurs. Grow on under glass until June.	January onwards	Greenhouse, or summer bedding plants in all but coldest areas. Seedlings will bloom some 18 months from sowing. Start tubers for pots into growth in greenhouse in February/March, for outdoors April. Indoors pot in 10–15 cm pots, outdoors plant 2.5 cm deep, 30 cm apart.
	SOFTWOOD CUTTINGS	Root 8 cm cuttings cut from the main stem at a leaf axil with a 'V' cut in a shaded propagator.	June/July	
	DIVISION (of tuber)	Replant in starting medium (peat) in greenhouse.	February/March as growth begins	
Begonia, Rhizomatous-rooted (Begonia rex)	LEAF CUTTINGS	Root in propagator at 16°C minimum, cutting leaf at main veins before placing flat down on compost. Separate and pot singly young plants when they have developed a few leaves.	During summer	Begonia rex is usually grown as a greenhouse/houseplant where it prefers a moist semi-shaded position. Grow seedlings on as for tuberous-rooted begonias, eventually potting singly in a peat-based potting compost and grow on in heated greenhouse.
	SEED (dust-like)	Sow in a temperature of 18–24°C.	March/April	
Belamcanda (B. chinensis = Blackberry Lily)	SEED	Sow singly in small pots in slight heat or cold frame and grow on in frames during the winter.	March / Late April	Half-hardy. Grow outdoors in mild areas only, elsewhere in greenhouse or frame. Flowers about 90 cm in height 3–4 years from sowing. Requires a rich sandy soil in full sun.
	DIVISION (of tuber)	Replant divisions 20 cm apart in growing positions.	August/September	

PLANT NAME	METHOD OF PROPAGATION	TREATMENT REQUIRED	TIME OF YEAR PROPAGATED	CULTURAL NOTES
Bessera (B. elegans = Coral Drops)	DIVISION (offsets; bulbs)	Grow offsets on in boxes of sandy soil in cold frame until large enough to plant out 8 cm deep, 20 cm apart.	June to September as foliage dies back	Half-hardy. Grow outdoors in full sun in mild areas, in large pots in greenhouse elsewhere. Grows 60 cm tall.
Bloomeria	DIVISION (offsets; corms)	Grow cormlets on in sandy compost in cold frame until large enough to plant out 10 cm deep, 15 cm apart.	After flowering when foliage dies back	Half-hardy. Grow in full sun outdoors in mild areas, in pots in greenhouse elsewhere. Makes good rock garden subject.
Bobartia	DIVISION (rhizomatous corm-like roots)	Replant where grown 10 cm deep, 15 cm apart outdoors or in pots in slightly heated greenhouse.	In autumn after flowering when pot grown plants are dried off	Half-hardy. Grow in warm sunny borders outdoors or 3–4 roots to a large pot in greenhouse. Height 45–60 cm.
Bomarea	DIVISION (fleshy tuberous roots)	Tease roots apart and replant 1 to a large pot or in greenhouse border.	March/April	Climbing plants for the warm greenhouse. Seedlings require a close moist atmosphere for their first 2 years. Pot singly when 5–8 cm high.
	SEED	Sow in a propagator at 18°C.	Autumn, as soon as ripe	
Bowiea	SEED	Sow in slight heat and grow on there transferring to cold frame to grow on in spring. Will take some years to reach flowering size.	Autumn or spring	Climbing plants for a sunny position against a south wall outdoors or in the greenhouse. Eventually forms a large bulb which will produce offsets.
Bravoa	DIVISION (of tuber: offsets)	Grow offsets on in boxes in greenhouse, frame or warm place outdoors. Offsets take up to 3 years to reach flowering size.	September/October	Half-hardy. Grow in pots in slightly heated greenhouse in all but the mildest areas.
Brodiaea (B. coccinea = Californian Firecracker/Crimson Satin Flower; B. hyacintha = Missouri Hyacinth)	SEED	Sow in cold frame or greenhouse and prick out into deep boxes to grow on.	As soon as ripe or in spring	Many species suitable for rock garden though some are tender and need greenhouse protection over winter. Plant out in warm sunny positions 15 cm deep, 15 cm apart, on sand.
	DIVISION (offsets; bulbs)	Grow on in deep boxes in frame for 2 years until flowering size is reached.	In autumn when flowers have died back	
Brunsvigia	SEED	Sow in a sandy compost in propagator and grow on in heated greenhouse. Will take 12–15 years to reach flowering size.	In spring	Plants produce large bulbs and require a warm greenhouse or conservatory in which to grow. Plant 1 bulb to a large pot and repot at 4 year intervals in spring. May be grown outdoors in very mild areas.
	DIVISION (offsets; bulbs)	Repot the occasional offsets in small pots and grow on potting annually.		
Bulbine (B. annua is an HHA growing 30 cm tall)	DIVISION (offsets; bulbs/fleshy tuberous roots)	Replant offsets in a warm sunny position 15 cm deep.	April/May	Half-hardy. Grow outside in mild areas only or lift and overwinter under frost-free glass.
Bulbinella (B. hookeri = Moari Onion)	DIVISION (of thick fleshy roots)	Replant offsets taken off when moving 51 cm apart in warm position.	In spring	Border plants growing from 30 cm to 1.2 m in height. Plant seedlings in final positions in spring.
	SEED	Sow in prepared bed outdoors.	April to June	
Bulbocodium (B. vernum = Spring Meadow Saffron)	DIVISION (offsets; bulbs)	Replant offsets 8 cm deep, 10 cm apart in dry sunny positions.	September/October	Good bulbs for rock garden or alpine house. Lift and divide every 2–3 years.
Caliphruria	DIVISION (natural layering by stolons)	Replant offsets 2–3 to a 13 cm pan. Ripen bulbs outdoors in sun during summer.	In autumn when repotting at 4 year intervals	Bulbs for a warm greenhouse with a minimum temperature of 10°C.

PLANT NAME	METHOD OF PROPAGATION	TREATMENT REQUIRED	TIME OF YEAR PROPAGATED	CULTURAL NOTES
Callipsyche (Sometimes included in Eucrosia)	DIVISION (offsets; bulbs)	Repot offsets in small pots and grow on in greenhouse.	Early autumn or early spring	Another genus needing a warm greenhouse minimum temperature 10°C.
Calochortus (C. amabilis = Golden Lanterns; C. elegans = Mariposa Lily)	DIVISION (offsets; bulbs and bulbils from leaf axils)	Grow offsets on in sandy soil in cold frame or greenhouse drying soil out each winter. Grow bulbils similarly, just pressing them into soil surface.	In autumn as leaves die back	Showy bulbs requiring extremely well-drained position in sun and protection from rain during autumn and winter. Plant 10 cm deep, 10 cm apart, in sand or raised bed. Do not disturb seedlings for 2 years then pot singly. All species can also be grown in pots permanently.
	SEED	Sow in sandy compost in slight heat and grow on under glass. Will flower 5–6 years from sowing.	As soon as ripe or in spring	
Calostemma (Australian Daffodil)	DIVISION (offsets; bulbs)	Grow offsets on in sandy compost in cold frame and plant out 15 cm deep, 30 cm apart when at flowering size.	Early spring	Grow outdoors in sun in mild areas only, elsewhere lift for the winter or grow in slight heat planting 2–3 bulbs per 15 cm pot.
Cammassia (C. quamash = Quamash)	DIVISION (offsets; bulbs)	Replant offsets 8 cm deep, 15 cm apart, in well-drained humus-rich soil.	In late summer when foliage has died back	Extremely hardy bulbs for borders where they grow 60–90 cm tall. Grow in sun or semi-shade. Seedlings will take 3–4 years to reach flowering size.
	SEED	Sow in cold frame and grow on there until bulbs form.	As soon as ripe or in spring	
Cardiocrinum (C. giganteum, syn. Lilium giganteum)	DIVISION (offsets; bulbs)	Parent bulb dies after flowering leaving a number of offsets which should be lifted and replanted 60 cm apart with nose just above the soil surface.	November	These plants require woodland conditions of semi-shade with a rich humus-filled soil. Offsets will flower 2 years from planting, seedlings 7–8. Protect from frost with a mulch of leaves. Plant out in final positions in June.
	SEED	Sow in cold frame or greenhouse and plant 8 cm apart in cold frame for first winter.	As soon as ripe	
Chasmanthe	DIVISION (offsets; corms)	Store detached small corms in frost-free place over winter and replant in May 10 cm deep, 20 cm apart.	Lift as foliage dies back	Border plants for sunny positions growing 60–90 cm tall. In very mild gardens lift every 4–5 years, elsewhere lift and store in frost-free place over winter.
	SEED	Sow in warm greenhouse or frame and grow on there for first year.	February/March	
Chionodoxa (Glory of the Snow)	DIVISION (offsets; bulbs)	Replant offsets in groups 5 cm deep in well-drained spot.	In summer when foliage dies back	Extremely hardy spring bulbs growing 8–23 cm in height. Ideal for rock gardens, troughs or alpine house.
	SEED	Sow thinly in cold frame and grow on there through winter.	June, as soon as ripe	
Chlidanthus	DIVISION (offsets; bulbs)	Grow offsets on in a sandy compost in cold frame or greenhouse.	In autumn when lifting for the winter	In warm areas grow at the base of a south wall, elsewhere 3 bulbs to 15 cm pot in greenhouse. Outdoors plant with nose above the surface 45 cm apart. Protect from rain and frost in winter or lift for storage, replanting April.
	SEED	Sow in greenhouse or frame and grow on there in small pots. Seedlings will flower 4–5 years from sowing.	In spring	
Clivia	DIVISION (offsets; fibrous bulb-like roots)	Remove offsets by pulling or cutting with a sharp knife. Repot in 13 cm pots and grow on in warm greenhouse.	Autumn or spring when plants start into growth	Greenhouse or in large tubs outdoors to house in minimum winter temperature of 10°C. C. miniata makes a good houseplant. Give plants plenty of water in growing season. Will flower 2 years from sowing.
	SEED	Sow singly in small pots in propagator at 21–24°C with humid atmosphere.	Spring or summer	

PLANT NAME	METHOD OF PROPAGATION	TREATMENT REQUIRED	TIME OF YEAR PROPAGATED	CULTURAL NOTES
Colchicum (C. autumnale = Meadow Saffron; the appearance of flowers before leaves has led to the nickname 'Naked Ladies')	DIVISION (offsets; corms) SEED	Replant offsets 10 cm deep, 15 cm apart, in well-drained spot in full sun. Sow in cold frame or prepared bed outdoors and transplant after 2 years to nursery beds.	July when foliage has died back As soon as ripe	Most flower in autumn the large leaves appearing in spring. Front of borders or among shrubs. Lift every 2–3 years. Topdress seedlings annually with potting compost. Flower 4–5 years from sowing.
Convallaria (C. majalis = Lily-of-the Valley)	DIVISION (rhizomes) SEED	Replant 5 cm deep, 10–15 cm apart, in moist ground with a north or east aspect. Sow in greenhouse or propagator and move 1-year-old seedlings to cold frames to grow on.	October August	Scented flowers good for cutting. Early flowering can be induced by 'forcing' boxed rhizomes in a heated greenhouse in March. Never let seedlings dry out, and shade from sun.
Cooperia	DIVISION (offsets: bulbs)	Store offsets in dry sand in frost-free place until spring. Grow on in frames planting 8 cm apart.	Autumn when lifting for winter storage	Grow outdoors in warm sunny spot in mild areas, elsewhere grow in greenhouse 3–4 to 15 cm pot. Outdoors plant 10 cm deep, 15 cm apart.
Crinum	DIVISION (offsets: bulbs) SEED (fleshy; just press into compost surface)	Grow offsets on in sandy compost in frame or greenhouse Sow in slight heat covering seed pan with glass until germination occurs. Prick out following summer into 15 cm pots to grow on.	March/April October	Rather tender. Choose warm sunny spot outdoors in mild areas. Protect from frost. Plant with neck of bulb just showing above soil. Can be grown in large tubs and housed in greenhouse over winter, or grow there permanently. Will take 2 years to flower from seed
Crocosmia (Montbretia)	DIVISION (offsets: corms) SEED	Replant divisions 10–13 cm deep, 30 cm apart, in rich soil in sun. Sow in cold frame or greenhouse and grow on there for 1 year undisturbed.	In spring as growth begins March/April	Border plants growing 60–90 cm in height. Lift and divide every 4–5 years. Some types may need frost protection in cold areas. Plant seedlings out in early summer.
Crocus (Crocus nudiflorus produces stolons)	DIVISION (offsets: corms) SEED	Replant offsets every 4 years 8 cm deep, 8–10 cm apart in fresh positions. Sow in cold frame and plant out after 2 years when corms have formed.	When foliage dies back July/August	Bright autumn/spring flowering bulbs for many situations in sunny positions in the garden or cold greenhouse. Seedlings will flower 3–4 years from sowing.
Cyclamen (hardy types; cyclamen europeum = Sowbread)	SEED DIVISION (of tuber)	Sow in pots and stratify bringing into cold frame in March. Grow seedlings on undisturbed for 1 year before placing outdoors. Replant according to species. Do not let tubers dry out.	December/January Early in autumn	Good plants for semi-shade around trees or in pans in an alpine house. Outdoors plant 13 mm deep in most cases. Most species will self-seed once established.
Cyclamen (greenhouse varieties; C. persicum, syn. C. latifolium = Persian Cyclamen)	SEED (soak in warm water for 24 hours before sowing)	Sow seed 13 mm deep and place in a temperature no higher than 15–18 °C. Prick out when two leaves have formed and grow on in slight heat.	In spring or autumn	Good greenhouse/houseplants requiring a minimum winter temperature of 4 °C. Flower 15 months from sowing. Shade seedlings and water carefully.
Cypella	DIVISION (offsets: bulbs)	Remove offsets on lifting and replant in boxes in a cold frame in spring. Plant out when bulbs reach flowering size 10 cm deep, 15 cm apart, in a warm sunny spot.	Lift in October and plant out in May	Grow outdoors in permanent position in warm areas only. Elsewhere store in frost-free place for the winter. Can also be raised from seed sown in frame as soon as ripe but germination is difficult.

PLANT NAME	METHOD OF PROPAGATION	TREATMENT REQUIRED	TIME OF YEAR PROPAGATED	CULTURAL NOTES
Cyrtanthus	DIVISION (offsets: bulbs)	Grow on in boxes in slightly heated greenhouse and pot 3–4 to a large pot. Offsets will flower 2 years from separation.	When repotting	A large number of species with differing habitats. Best grown in slight heat but in warm areas some can be flowered outside in a warm sunny spot.
Dierama (Wand Flowers/Angels' Fishing Rods)	DIVISION (corm-like roots)	Cut roots into corm-shaped sections and replant 8 cm deep, 45 cm apart.	April	Border plants growing from 0.6 m to 1.8 m in height. Prefer a moist lime-free soil in sun. Protect from frost in cold areas. Overwinter seedlings in cold frames until large enough for planting out.
	SEED	Sow fresh seed in a cold frame and overwinter there or sow purchased seeds in spring. Grow on in pots.	October	
Dipcadi	DIVISION (offsets: bulbs)	Detach offsets and grow on in greenhouse. Offsets will flower 2 years from separation.	In autumn when lifting for winter	Cultivate outdoors in mild areas and in pots in a greenhouse elsewhere. Outdoors lift in autumn and store in dry frost-free place. Most grow 90 cm in height but *D. serotinum* at 23 cm makes a good rock garden plant.
	SEED	Sow in a warm greenhouse and grow on there to flowering size.	In spring	
Disporum (Fairybells)	DIVISION (Rhizomatous roots)	Lift and divide roots with a knife and replant 8 cm deep, 38 cm apart.	In October when leaves die back	Herbaceous border plants requiring woodland conditions of semi-shade and moist leafy soil.
	SEED	Sow in cold frame and grow on there for 18 months before planting out.	July	
Drimia	DIVISION (offsets: bulbs)	Remove offsets when lifting or repotting. Grow on in boxes in a greenhouse and pot 4–5 to a 13 cm pot after 3 years.	In autumn after flowering	Half-hardy. Grow under glass in all but very mild areas planting in a sunny spot 10 cm deep, 20 cm apart.
Endymion (E. non-scripta = English Bluebell or Wood Hyacinth)	DIVISION (offsets: bulbs)	Lift and replant bulbils immediately outdoors in sun or semishade planting 8 cm deep, 15 cm apart.	In late summer when foliage has died down	Species include both English and Spanish Bluebells previously included under *Scilla* and *Hyacinthus*. Established colonies will self-seed, or seed can be sown among shrubs or trees where the soil is bare.
	SEED	Sow in cold frame and grow on for 15 months before planting out.	July when seed is ripe	
Eranthis (Winter Aconite)	DIVISION (of tubers)	Lift older tubers and divide several times with a knife. Replant 5–8 cm deep, 10–15 cm apart.	October	Carpeting plants for semi-shade beneath shrubs etc. Select an odd tuber each year for division as disturbance causes loss of flowers. Seedlings will take 4 years to reach flowering size. Grow on in cold frame for 18 months after transplanting.
	SEED (slow germination)	Sow in cold frame. Seedlings appear in March/April and remain green for a few weeks only. Transplant when growth restarts.	July	
Erythronium (E. dens-canis = Dog's Tooth Violet; E. revolutum = Trout Lily)	DIVISION (offsets: bulb-like tubers)	Older plants can be lifted and divided. Replant immediately 8 cm deep, 15 cm apart. In dry seasons water well until established.	May, after flowering but before leaves die back	These plants dislike disturbance and are best left to increase naturally. Plant in moist leafy soil slight shade. Seedlings will flower 4–5 years from sowing.
(E. albidum and E. americanum are increased from stolons)	SEED (surface sow)	Sow in cold frame and keep compost moist and shaded from sun. Grow on and plant out in third year.	July when seed is ripe	
Eucharis (E. amazonica = Amazon Lily; E. candida bears stolons)	DIVISION (offsets: bulbs)	Remove small bulbils from around older bulbs and pot on singly in 13 cm pots.	March or at the end of a 'rest' period	Plants for a warm greenhouse with a minimum winter temperature of 18 °C.

PLANT NAME	METHOD OF PROPAGATION	TREATMENT REQUIRED	TIME OF YEAR PROPAGATED	CULTURAL NOTES
Eucomis (E. bicolor = Pineapple Flower)	DIVISION (offsets: bulbs)	Remove offsets and grow on in cold greenhouse or frame planting 1 bulb to a 13 cm pot.	November when lifting or March when repotting	Hardy in mild areas only elsewhere grow in a cold greenhouse. Outdoors choose a warm spot and plant out 20 cm deep 30 cm apart. Protect from frost during winter.
	SEED (slow germination)	Sow in a gritty compost in a warm greenhouse and grow on after 1 year in 8 cm pots.	In spring	
Eucrosia	DIVISION (offsets: bulbs)	Repot bulbs every few years and remove offsets. Grow these on in boxes until flowering size is reached. Pot 3 or 4 to an 18 cm pot.	In autumn	After 8 weeks in a cold frame to form roots, these bulbs require a minimum temperature of 10 °C during winter.
Eurycles	DIVISION (offsets: bulbs)	Repot every 5 years and remove offsets. Grow on in boxes until at flowering size then pot singly in 13 cm pots.	In October	Plants for a warm greenhouse requiring a minimum temperature of 18 °C when growing and 16 °C during their resting period.
Ferraria	DIVISION (offsets: corms)	Remove cormlets when lifting or repotting. Store in a frost-free place over winter if necessary and grow on in boxes in a greenhouse for two years before potting on.	In autumn when lifting or in spring	Grow outdoors only in warm areas where they need much sun. Plant out 8 cm deep on sand 15 cm apart. Indoors plant 3–4 corms to a 13 cm pot. Maintain a winter temperature of 7–10 °C.
Freesia	DIVISION (offsets: corms)	Propagate selected strains from cormlets. Grow on as for seedlings overwintering under frost-free glass.	In spring	Only hardy in mild areas, elsewhere they require a minimum winter temperature of 9 °C. Place outdoors through the summer and house in September. Grow as pot plants or cut flowers.
	SEED (soak 24 hrs in warm water before sowing)	Sow in a propagator at 13–18 °C. 6 seeds to a 13 cm pot. Do not transplant.	March to June	
Fritillaria (F. imperialis = Crown Imperial; F. meleagris = Chequered Daffodil/Snakeshead Fritillary)	DIVISION (offsets: bulbs)	Detach offsets when lifting and replant immediately handling with care.	October	Crown Imperials grow 1 m in height. Plant 15 cm deep on their side and 45 cm apart in sun. Others prefer semi-shade in moist soil. Plant 8 cm deep, 10–15 cm apart.
	SEED	Sow in cold frame or greenhouse. Transplant 18 months from sowing and grow on in pots or boxes. Will flower 4 years from sowing.	August	
Gagea (G. lutea = Yellow Star of Bethlehem)	DIVISION (offsets: bulbs)	Lift and separate older groups. Replant 8 cm deep, 10 cm apart.	In autumn	Plants for semi-shade in moist leafy soil. Produce bulbils in leaf axils which can be grown on as offsets.
Galanthus (Snowdrop)	DIVISION (offsets: bulbs)	Divide older groups before foliage has died down. Replant immediately 8–10 cm deep, 10 cm apart.	March/April	Extremely hardy. Plant in rock gardens or border. Can be grown in pots to flower indoors but give them little or no heat. Grow on in outdoor beds in semi-shade 5 cm apart.
	SEEDS	Sow in cold frame and keep compost moist. Transplant 12–15 months later.	June, as soon as ripe	
Galaxia	DIVISION (offsets: corms)	Cormlets form on leaf axils and can be removed when ripe. Grow on 3–4 corms to a 15 cm pan.	Late summer or autumn	Low-growing plants for a slightly heated greenhouse or outdoors in mild areas 10 cm deep 10 cm apart in sunny warm position.
Galtonia (Summer Hyacinth; G. candicans = Spire Lily)	DIVISION (offsets: bulbs)	Lift and detach offsets and replant in groups 15 cm deep, 30 cm apart.	April	Border plants growing from 38 cm to 1.2 m in height. Hardy in all but the coldest areas. Protect from frost with a mulch of leaves.
	SEED	Sow in cold frame and grow on there in pots. Plant out after 3 years. Will flower in about 5 years.	April	

PLANT NAME	METHOD OF PROPAGATION	TREATMENT REQUIRED	TIME OF YEAR PROPAGATED	CULTURAL NOTES
Geissorhiza	DIVISION (offsets: corms)	Detach cormlets and store in dry sand in a frost-free place overwinter. Grow on in boxes and protect from frost in winter.	In autumn when lifting for winter storage	Hardy in mild areas where they are planted 10 cm deep, 15 cm apart, at the foot of a warm wall. Elsewhere grow 4 corms to a 15 cm pot in greenhouses.
Gethyllis	DIVISION (offsets: bulbs)	Remove offsets when repotting and grow on in a sandy compost. Will flower 2 years from separation.	April	Crocus-like plants requiring cold greenhouse protection in all but mild areas. Lift and store in frost-free place during winter.
Gladiolus (Sword Lily/Corn Flag)	DIVISION (offsets: corms)	Remove cormlets from new corms and plant in deep boxes. Place in a cold frame and grow on as for seedlings.	In autumn when lifting	Ideal border plants or cut flower. Grow in sun or semi-shade planting corms 8–13 cm deep, 20–25 cm apart. Seedlings and cormlets will each take 2 years to reach flowering size. Lift and store in a frost-free place during winter. Plant young corms out in drills 5 cm deep, 30 cm apart, for their 2nd year.
	SEED (cover boxes with glass and remove when seedlings are 2.5 cm high)	Sow in deep boxes in a temperature of 13–16°C. Harden off and place outside for the summer. Corms will have formed by autumn.	March	
Gloriosa (Flame Lily)	DIVISION (offsets: bulb-like tubers)	Remove offsets when repotting and grow on singly in small pots.	February to May	Climbing plants for a warm greenhouse with a minimum temperature of 15°C during growing season, rather less when dried out for its winter rest.
	SEED	Sow singly in small pots in a temperature of 20–24°C. Pot on as necessary.	In spring	
Griffinia	DIVISION (offsets: bulbs)	Remove offsets when repotting and press singly into sandy compost in 8 cm pots.	In spring after flowering	These bulbs require a slightly heated greenhouse with a minimum temperature of 10°C. Repot every 3 years. Seedlings will flower 5 years from sowing.
	SEED (surface sow)	Sow in a temperature of 18–24°C. Move to 8 cm pots in following April.	June	
Habranthus	DIVISION (offsets: bulbs)	Remove offsets when repotting and grow on singly in 8 cm pots.	In September or March depending on flowering time	Cold greenhouse protection is needed in most areas. Add a little old mortar rubble to the compost and pot 1 large bulb to 15 cm pot leaving top of bulb exposed. Water carefully in winter.
	SEED	Sow singly in small pots in a temperature of 16°C. Grow on in slight warmth.	As soon as ripe	
Haemanthus (Blood Flower/Red Cape Lily)	DIVISION (offsets: bulbs)	Remove offsets when repotting and pot singly in 8 cm pots. Dry off for winter and start into growth in March in a temperature of 13°C.	In autumn when leaves die back	Most species require slight heat, though *H. katharinae* can be grown outdoors in tubs during summer.
Herbertia	DIVISION (offsets: corms)	Detach cormlets when lifting or de-potting for frost-free winter protection. Grow on in boxes in spring and dry out each winter. Will take 3–4 years to flower.	Autumn	Most species need slight heat, but *H. pulchella* will grow in sunny borders outdoors. Plant out in April 10 cm deep, 15 cm apart. Indoors plant 4–5 corms to a 15 cm pan.
Hermodactylus (H. tuberosus, syn. Iris tuberosus = Snake's Head Iris	DIVISION (of tubers)	Lift and divide tubers when overcrowded.	In summer as leaves die back	Plant in full sun in well-drained lime-rich soil 8 cm deep, 15 cm apart.
Hesperantha	DIVISION (offsets: corms)	Detach cormlets when lifting for frost-free winter storage. Grow on in pans under glass, 6 corms to 15 cm pan.	In autumn	Require greenhouse protection in all but very mild areas. Plant out in April 10 cm deep, 10 cm apart. Choose a warm spot.

PLANT NAME	METHOD OF PROPAGATION	TREATMENT REQUIRED	TIME OF YEAR PROPAGATED	CULTURAL NOTES
Hexaglottis	DIVISION (offsets: corms)	Remove cormlets and store in boxes of dry sand. Replant in spring in boxes 2.5 cm apart pressing cormlets into compost surface and watering when growth begins.	In autumn when lifting for frost-free winter storage	Grow outdoors in sunny positions planting 8 cm deep, 15 cm apart. Indoors, plant 4–5 corms to 15 cm pot. Flowers have an unpleasant scent.
Hippeastrum *(also known as Amaryllis hippeastrum; H. equestre = Barbados Lily)*	DIVISION (offsets: bulbs)	Remove offsets when repotting and grow on in boxes for 1 year before potting singly in small pots.	January	Most species make good house/greenhouse plants. *H. pratense* can be grown outdoors in mild areas. Indoors plant one large bulb to a 15 cm pot leaving top of bulb exposed. Seedlings will flower in 3–4 years.
	SEED	Sow in a temperature of 16–18°C, one seed to an 8 cm pot, and grow on in greenhouse.	As soon as ripe	
Homeria *(H. collina miniata = Red Tulip)*	DIVISION (offsets: corms and bulbils formed in leaf axils)	Detach cormlets when lifting and grow on in boxes in slight heat. Bulbils can be taken off to be grown on as for cormlets.	In autumn / In late summer	Grow outdoors in frost-free gardens in sunny places 15 cm deep 15 cm apart. Elsewhere grow in slight heat 6 corms to a 15 cm pot.
Homoglossum	DIVISION (offsets: corms)	Remove cormlets when repotting and grow on in sandy compost. Will flower 4 years from separation.	In autumn	Best grown in slight heat as many flower in winter. Plant 4 or 5 corms to a 15 cm pot.
Hyacinthus *(H. orientalis gave rise to the large-flowered 'Dutch' hyacinths)*	DIVISION (offsets: bulbs)	Cut mature bulbs 13 cm deep with a cross across the root base before planting. Offsets will form at cuts to be removed the following summer. Will take 3 to 6 years to reach flowering size.	September	Large flowered types make good pot plants or bedding plants, others use in rock garden. Slow to propagate naturally. Grow offsets on in rich well-drained yet moist soil in full sun. Remove flowers as they appear from mother bulbs.
Hymenocallis *(Ismene)*	DIVISION (offsets: bulbs)	Remove offsets on repotting every 3–4 years. Grow on in small pots.	In spring	Plant hardy species 15 cm deep, 23 cm apart outdoors, others 1 bulb to a suitable sized pot in a slightly heated greenhouse. *H. narcissiflora* is good in tubs outdoors in summer. Dry young plants off for winter.
	SEED	Sow singly in small pots in a temperature of 16–18°C. Pot annually until flowering size is reached.	April	
Hypoxis *(H. hirsuta = Yellow Star Grass)*	DIVISION (offsets: bulbs)	Remove offsets when repotting and grow on in small pots.	September	Plants requiring slight heat. Plant 2.5 cm deep, 2–3 to a 15 cm pot.
Ipheion	DIVISION (offsets: bulbs)	Remove offsets when lifting or repotting. Grow on in boxes and protect from frost in winter. Will flower 4 years from separation.	October	Grow outside in all but very cold areas in full sun or semi-shade amongst shrubs. Plant 10 cm deep, 10 cm apart or 4–5 bulbs to a 13 cm pot indoors.
Iris *(bulbous types)*	DIVISION (offsets)	Lift and divide established groups, drying them off slowly in the sun to ripen them, and replanting in September or October.	As leaves die back	Plant bulbs of the *reticulata* type 5 cm deep, 5 cm apart, or 5–6 to a 13 cm pan, *xiphium* types 8–10 cm deep, 15 cm apart.
Ixia *(African Corn Lily)*	DIVISION (offsets: corms)	Remove cormlets when lifting or repotting. Grow on in boxes planting 13 mm deep, 5 cm apart in frame or greenhouse.	In autumn	Hardy in mild areas in warm spots planted 10 cm deep, 15 cm apart. Elsewhere grow in slight heat 6–8 corms to 13 cm pot. Dry corms out after flowering for 2 months. Seedlings will flower in 4 years.
	SEED	Sow in shaded greenhouse or frame and transplant after 2 years to deep boxes to grow on.	In spring	
Ixiolirion *(Blue Altai Lily)*	DIVISION (offsets: bulbs)	Remove offsets after flowering and grow on singly in 8 cm pots for first 2 years. Will flower 4 years from separation.	Late summer/early autumn	Plant out 8 cm deep, 15 cm apart, in warm positions and protect from frosts and too much winter wet in cold areas.

PLANT NAME	METHOD OF PROPAGATION	TREATMENT REQUIRED	TIME OF YEAR PROPAGATED	CULTURAL NOTES
Lachenalia *(Cape Cowslip/Leopard Lily)*	DIVISION (offsets: bulbs)	Repot offsets singly in 8 cm pots and place out in plunge bed outdoors for 2 months as for older bulbs.	In autumn	Hardy in the mildest areas but elsewhere grow 5–6 bulbs to a 15 cm pan. Outdoors plant 10 cm deep, 20 cm apart. Seedlings take 3–4 years to reach flowering size.
	SEED	Sow in a temperature of 16–18°C. Grow on in 8 cm pots in slight heat.	March/April	
Lapeyrousia	DIVISION (offsets: corms)	Remove offsets when repotting. Grow on in boxes.	In spring	Best grown in slight heat in other than frost-free gardens. Under glass plant 6 corms to a 13 cm pan.
	SEED	Sow in slight heat in boxes and dry out gradually for winter.	April	
Leucocoryne *(L. ixiodes = Glory of the Sun)*	DIVISION (offsets: bulbs)	Remove offsets when repotting. Plant 2.5 cm deep, 5 cm apart, in boxes to grow on. Flower in 2 years.	In autumn	Good pot plants or cut flowers requiring slight heat in winter. Plant 4–5 to 15 cm pot.
Leucojum *(L. vernum = Spring Snowflake; L. aestivum = Summer Snowflake*	DIVISION (offsets: bulbs)	Divide older groups and replant at once 8 cm deep, 15 cm apart. Grow small bulbs on in boxes for 2 years.	Soon after flowering	Similar in appearance and culture as *Galanthus* though some may need glass protection in cold areas. Seedlings will flower in 4–5 years. Grow on in boxes of sandy compost.
	SEED	Place outdoors to stratify and move to a frame in March.	September	
Libertia	DIVISION (rhizomatous roots)	Divide as growth begins replanting 8 cm deep, 15 cm apart, in peaty soil in sun.	Early spring	Hardy in mild areas, elsewhere grow in slight heat 1–2 rhizomes to a 15 cm pot.
Lilium *(L. candidum = Madonna Lily; L. tigrinum = Tiger Lily; L. regale = Regal Lily)*	DIVISION (of roots)	Lift and divide overcrowded clumps using hands to make the division to avoid bulb damage.	September/October	Many species and varieties ideal for garden and/or greenhouse culture. Many grow best with their roots in shade and their tops in full sun. Grow in moist but well-drained soil. When planted leave bulbils undisturbed for 2 years then plant out in 3rd autumn. With scales, bulbils will form at their base in about 3 months. In spring treat as for leaf axil bulbils. Remove any flowers that form in first 2 years. Transplant seedlings when 2.5 cm high and grow on in 8 cm pots in a cold frame for first few years of growth.
	DIVISION (offsets: bulblets)	Bulblets form in leaf axils in some species. Remove these and plant 2.5 cm deep, 5 cm apart, in drills in prepared bed outdoors or in deep boxes in cold frame.	In autumn when ripe	
	DIVISION (offsets: scales)	Carefully remove scales from bulbs and plant upright in moist seed compost in a propagator at 18°C.	September/October	
	SEED (variable germination rate according to the species)	Sow 2.5 cm deep, 2.5 cm apart, in cold frame or sow in a raised bed outdoors in spring.	In autumn when ripe	
Liriopsis *(Plagiolirion)*	DIVISION (offsets: bulbs)	Remove offsets when repotting every 2 years. Grow on in 8 cm pots where they will flower in 2 years.	Early in spring	Bulbs requiring a minimum temperature of 13°C. Plant 2–3 bulbs to a 13 cm pot.
Littonia	DIVISION (of tubers)	Divide with a knife and replant singly in 10 cm pots covering tubers with 5 cm of compost.	In spring	Climbing plants requiring slight heat and similar treatment as that for *Gloriosa*.
Lloydia *(L. serotina = Snowdon Lily)*	DIVISION (offsets: bulbs)	Remove offsets from older bulbs and replant 5 cm deep, 15–20 cm apart.	September/October	Hardy bulbs for the rock garden in semi-shade in moist but well-drained soil.
Lycoris *(L. aurea = Golden Spider Lily)*	DIVISION (offsets: bulbs)	Remove offsets when repotting every 4–5 years and grow on in 8 cm pots in the greenhouse.	August	Grow outdoors in mild areas, elsewhere in slight heat 1 bulb to a 13 cm pot. Offsets will flower in 4 years as will seedlings. Pot to 13 cm pots after 2 years' growth.
	SEED (surface sow)	Sow in slight heat and move singly to 8 cm pots about 18 months from sowing.	April	

PLANT NAME	METHOD OF PROPAGATION	TREATMENT REQUIRED	TIME OF YEAR PROPAGATED	CULTURAL NOTES
Massonia	DIVISION (offsets: bulbs)	Remove offsets when repotting and grow on in gritty compost. Offsets will flower in about 3 years.	In autumn	Place pots out in plunge beds in autumn and house in slight heat in November. Grow 3–4 bulbs to a 15 cm pot.
Medeola (Cucumber Root)	DIVISION (rhizomatous roots)	Lift and divide older groups replanting just below soil surface 30 cm apart.	In autumn as leaves die back or in spring	Border plants 30–45 cm in height. Grow in semi-shade in moist leafy soil.
Melasphaerula	DIVISION (offsets: corms)	Remove offsets when repotting. Grow on for 2 years in pans or boxes.	In late summer when foliage dies back	Plants best grown on in slight heat planting about 6 corms to a 15 cm pot.
Merendera (M. montana = Pyrenees Meadow Saffron)	DIVISION (offsets: corms)	Lift older groups and remove small cormlets. Grow on in boxes in cold frame.	July	Suitable for warm rock gardens or the alpine house 6 corms to a 15 cm pan of gritty compost. Outdoors plant 5 cm deep, 13–15 cm apart. Disturb as little as possible.
	SEED	Sow in cold frame and topdress annually with decayed manure. Plant out or pot after 3 years.	In spring	
Moraea (Butterfly Iris)	DIVISION (offsets: corms)	Detach cormlets when repotting and grow on in boxes. Will flower in 2 years.	September	Grow *M. spathulata* (syn. *M. spathacea*) outdoors in mild areas, elsewhere in slight heat as for other species planting 5–6 corms to a 15 cm pot. Seedlings flower 4–5 years from sowing.
	SEED	Sow in greenhouse and transplant 2.5 cm apart after 2 years.	June	
Muscari (Grape Hyacinths)	DIVISION (offsets: bulbs)	Lift older clumps and replant offsets in groups 8 cm deep, 10–13 cm apart, or grow on in boxes.	September	Good carpeters for sunny spots in the shrubbery or in pots in the cold greenhouse. In pots grow 5–6 bulbs to a 15 cm pot. Seedlings will flower 4–5 years from sowing.
	SEED	Sow in cold frame and grow on for 2 years before planting out.	April	
Narcissus (includes Daffodils)	DIVISION (offsets: bulbs)	Lift large flowered types every 4–5 years and replant largest offsets 13 cm deep, 15–20 cm apart.	In summer when leaves die back	Lift species in May wherever overcrowded. Divide into groups of 3 or 4 and replant at once the same depth as before.
Nemastylis	DIVISION (offsets: corms)	Remove cormlets on lifting or repotting and grow on in boxes in frames. Offsets will flower in 3–4 years. Place outdoors during summer.	October	Grow outdoors in mild areas; elsewhere in a little warmth in winter. Plant out 10 cm deep, 15 cm apart, or 3–4 corms to a 15 cm pot.
Nerine	DIVISION (offsets: bulbs)	Repot every 4 years and grow offsets on for 3 in boxes in the greenhouse.	July/early August	Grow outdoors in mild areas or sheltered places, *N. bowdenii* being most suitable. Others grow in a minimum winter heat of 7°C, one to a pot. Outdoors plant 8 cm deep, 23 cm apart. Dry out for a few months when leaves die back.
	SEED (surface sow)	Sow on a gritty compost in the greenhouse at 13–16°C. Transplant to 8 cm pots in autumn. Will flower some 3 or 4 years from sowing.	During spring	
Nomocharis (N. oxypetala, syn. Lilium oxypetalum)	SEED	Sow in pots in cold frame keeping compost moist and frame well ventilated. Plant out 10 cm deep, 23 cm apart after 2 years, disturbing as little as possible.	April	Seed in generally available for these rarely cultivated plants. Require cold in winter and as much sun as possible in summer.
Notholirion	DIVISION (offsets: bulbs)	Bulbs die after flowering but bulbils form at old bulb's base. Grow these on in boxes placing them outdoors in summer. Will flower 4 years from separation.	August/September	Only suitable to outdoor culture in dry, mild areas, elsewhere under frost-free glass 1 bulb to a 13 cm pot. Outdoors plant 10 cm deep, 30 cm apart.

PLANT NAME	METHOD OF PROPAGATION	TREATMENT REQUIRED	TIME OF YEAR PROPAGATED	CULTURAL NOTES
Nothoscordum	DIVISION (offsets: bulbs)	Lift older groups and replant offsets in boxes and grow on in cold frame.	August/September	Plant in groups among shrubs 8 cm deep, 15 cm apart. Offsets will flower some 3 years from separation.
Ornithogalum (O. umbellatum = Star of Bethlehem; O. thyrsoides = Chincherinchee	DIVISION (offsets: bulbs)	Lift older clumps and grow offsets on in boxes for 2 years before planting out.	In autumn	Plant in groups in borders or rock garden. Plant in sun or semi-shade 8–10 cm deep, 13–20 cm apart. Most species are hardy. Grow on outside in boxes for a year or so before planting out.
	SEED	Sow shallowly in pans in a cold frame and topdress with compost after the first season's growth.	In spring	
Oxalis (Sorrel)	DIVISION (of bulb-like rhizomes)	Lift and divide older plants replanting sections just below soil surface 15 cm apart.	In spring	Grow hardiest species in the rock garden, others 1 to a 8–10 cm pan in the alpine house. Seedlings will flower in 2–3 years.
	SEED	Sow in greenhouse or frame and pot on after 1 year.	In spring	
Pamianthe	DIVISION (offsets: bulbs)	Repot every 3 years. Remove offsets and grow on in small pots.	September	Requires a winter minimum temperature of 10°C. Move to larger pots each year, 1 to 13 cm pot when at flowering size.
Pancratium	DIVISION (offsets: bulbs)	Detach offsets when lifting or repotting and grow on in 8 cm pot. Grow on for 2 years then plant out 15 cm deep, 30 cm apart or 1 to a 13 cm pot.	October	Most species require some heat though those such as *P. illyricum* can be grown outdoors in mild areas. Protect from frost and excessive moisture in winter.
Paradisea (St Bruno's Lily)	DIVISION (of tubers)	Lift and divide established plants replanting at once 5 cm deep, 23 cm apart.	September/October or in spring	Extremely hardy border or rock garden subject growing 60 cm in height. Can also be raised from seed sown in prepared beds outdoors.
Phaedranassa	DIVISION (offsets; bulbs)	Remove offsets when repotting and grow on in 8 cm pots. Will flower in 2 years.	October	Requires only slight heat to give frost protection. Plant 1 bulb to a 13 cm pot.
Polianthes (Tuberose)	DIVISION (offsets; bulb-like tubers)	Offsets can be grown on to flowering but require more sunshine than is usual in the British Isles to ripen them.	In spring	Greenhouse plants for cut flowers. Tubers are usually discarded after flowering. Plant tubers in spring 1 to 10 cm pot.
Polygonatum (Solomon's Seal)	DIVISION (of rhizomes)	Lift and divide older plants with a knife and replant just below soil surface 30 cm apart.	November	Border plants for a moist leafy soil in semi-shade.
Puschkinia (Striped Squill)	DIVISION (offsets; bulbs)	Lift congested clumps remove smaller offsets and grow on in boxes for 2 years before planting out.	October	Plant in rock garden or among shrubs 8 cm apart 8 cm deep, or grow 6–7 to a 13 cm pot planting out after flowering.
Ranunculus (tuberous-rooted types)	DIVISION (offsets; claw-like tubers)	Remove offsets when lifting for winter storage. Grow on in boxes in spring.	When foliage dies down	Plant out in borders claws down 8 cm deep, 10–13 cm apart. Lift in autumn for frost-free winter storage. Seedlings will flower 2–3 years from sowing. Also grown in pots in greenhouse.
	SEED	Sow in cold frame or greenhouse and prick out 2.5 cm apart after 1 year and grow on in cold frame.	April	
Rechsteineria (formerly in Gesneria)	DIVISION (of tuber)	Cut tuber into sections with a knife and pot singly in 13 cm pot.	Start tubers in peat in March	Pot plants requiring a warm greenhouse with minimum summer temperature of 18°C and 13°C in winter. Grow cuttings and seedlings on in 8 cm pots. Seedling will flower 6–7 months from sowing.
	SOFTWOOD CUTTINGS	Take nodal cuttings or with piece of tuber attached. Root in a warm propagator.	March/April	
	SEED (surface sow)	Sow in a temperature of 18–24°C.	In spring or summer	

PLANT NAME	METHOD OF PROPAGATION	TREATMENT REQUIRED	TIME OF YEAR PROPAGATED	CULTURAL NOTES
Reineckea	DIVISION (of rhizomes)	Lift established plants and divide with a knife.	In autumn	Border plants growing 30 cm in height. Plant 8 cm deep, 30 cm apart.
Rigidella	DIVISION (offsets; corms)	Remove offsets when repotting and grow on in boxes. Offsets will flower in 2 years.	In autumn after flowering	Grow in pots under glass in a minimum winter temperature of 7°C. *R. flammea* grows 90 cm or more in height.
Romulea	SEED	Sow in slight heat in greenhouse or frame keeping compost moist. Grow on undisturbed for 2 years when they should begin to flower. Keep fairly dry through the winter.	April	Crocus-like corm forming plants for the rock garden. Plant in October 8 cm deep, 8 cm apart. Protect from frost in winter. Can also be planted 5–6 corms to a 13 cm pan in greenhouse.
Sandersonia (Christmas Bells)	DIVISION (of tubers)	Lift as growth begins and divide with a knife. Outside plant against a warm wall 5 cm deep.	In spring	Climbing plants similar to *Gloriosa*. Grow 1 to a 10 cm pot or grow outside and lift for winter storage.
Schizostylis	DIVISION (of rhizome)	Lift and divide older plants and replant 10 cm deep, 8 cm apart.	April	Grow outdoors only in areas with warm autumns. Elsewhere grow in deep boxes placed out in summer and housed in slight heat in autumn.
Scilla (Squill; *S. verna* = Spring Squill)	DIVISION (offsets; bulbs)	Lift and divide every 4–5 years and replant small bulbs in drills planting 5 cm deep, 2.5 cm apart. Grow on for 2 years.	Spring or summer when foliage dies back.	Most species are very hardy. Grow in sun or semi-shade in rock gardens, troughs or 5–6 to a 13 cm pan. Plant out 8 cm deep 10 cm apart. Grow seedlings on as for offsets. Will flower in 4–5 years.
	SEED	Sow in boxes in cold frame to germinate then move to shady spot outdoors.	In spring	
Sinningia (*S. speciosa* = Gloxinia)	DIVISION (of tuber)	Divide tubers with a knife when growth has begun and pot singly in 13 cm pots.	April/May	Require a heated greenhouse with a minimum temperature of 16°C. They make good houseplants when in flower. Gradually dry out for the winter protecting the tubers from frost. Move seedlings to 8 cm pots when large enough.
	LEAF CUTTINGS	Use mature leaves with stalk and root in propagator at 21°C.	During summer	
	SEED	Sow in a temperature of 18–24°C. Will flower in late summer.	January/February	
Smithiantha (Naegelia)	DIVISION (of rhizome)	Divide rhizomes and replant immediately 2.5 cm deep in 8 cm pots.	In spring	Achimene-like plants requiring a greenhouse with a minimum summer temperature of 13°C and 7°C in winter.
	LEAF CUTTINGS	Use mature leaves with leaf stalk attached. Root in humid, shady propagator at 16–18°C.	July	
Sparaxis (Harlequin Flower)	DIVISION (offsets; corms)	Lift when dormant and remove offsets to grow on in boxes in slight heat. Will flower in 2 years.	September	Grow under glass in slight warmth except in very mild areas where they can be planted out 10 cm deep, 10 cm apart.
Sprekelia (Jacobean or Aztec Lily)	DIVISION (offsets; bulbs)	Repot every 3–4 years growing small offsets on in boxes, larger in 8 cm pots in slight heat. Offsets will flower in 3–4 years.	September	Plant outdoors 10 cm deep, 15 cm apart, in very mild areas, elsewhere 1 bulb to 13 cm pot under heated glass.
Stenomesson	DIVISION (offsets; bulbs)	Remove offsets when repotting every 3–4 years and grow on in a warm place. Will flower in about 2 years.	November	Grow in a warm greenhouse planting 1 bulb to a 13 cm pot.

PLANT NAME	METHOD OF PROPAGATION	TREATMENT REQUIRED	TIME OF YEAR PROPAGATED	CULTURAL NOTES
Sternbergia (S. lutea = Lily of the Field)	DIVISION (offsets; bulbs)	Lift established groups, remove offsets and grow on in 8 cm pots in slight heat. Plant out 10 cm deep, 15 cm apart after 2 years.	When leaves die down	Require an open sun-baked position in a lime-rich soil in areas with mild winters. Offsets will flower some 4 years from separation.
Streptanthera	DIVISION (offsets; corms)	Repot each year and grow offsets on in a frost-free greenhouse. Grow on for 2 years before potting or planting out.	October	Ixia-like plants for outside cultivation in mild areas planting 10 cm deep, 10 cm apart, or grow in pots in slight heat.
Synnotia	DIVISION (offsets; corms)	Detach offsets when repotting and grow on in boxes for 2 years.	October	Grow in frost-free greenhouse planting 3–4 corms to a 13 cm pot.
Syringodea	DIVISION (offsets; corms)	Remove cormlets when lifting or repotting and grow on in boxes in slight warmth. Offsets will flower in 2 years.	In late autumn	Grow in greenhouse or warm spot in rock garden lifting for the winter. Plant out in April 5 cm deep, 8 cm apart, or 5–6 corms to a 15 cm pan.
Tecophilaea (Chilean Crocus)	DIVISION (offsets; corms)	Lift or repot every 3–4 years growing offsets on in boxes in the greenhouse or frame protecting from frost in winter.	In autumn	Require cold house protection in most areas. Plant out in October 15 cm deep, 10 cm apart, or 5–6 corms to a 15 cm pan. Seedlings will flower in about 4 years.
	SEED	Sow in greenhouse or frame and grow on as for offsets.	May	
Tigridia (Mexican Shell Flower/Tiger Flower)	DIVISION (offsets; corms)	Detach offsets when lifting or repotting and grow on in boxes in the greenhouse.	In autumn	Grow outdoors in mild areas planting 10 cm deep, 15 cm apart. Elsewhere grow 5–6 to a 13 cm pot in slight heat. Seedlings flower after 5 years of growth.
	SEED	Sow in cold frame and pot on after 3 years. Protect from frost in winter.		
Trillium (Wood Lilies; T. grandiflorum = Wake Lily or Snow Trillium)	DIVISION (of rhizome)	Older plants can be lifted and divided, replanting pieces at once 10 cm deep, 15–20 cm apart.	In autumn	Extremely hardy plants for a moist leafy soil in semi-shaded borders. Seedlings will take 5–6 years to grow to flowering size. Grow on in pots outdoors or in cold frame.
	SEED (slow to germinate)	Sow in cold frame and keep compost moist.	In late summer when ripe	
Tritonia	DIVISION (offsets; corms)	Corms form on the rhizomes and can be removed on lifting or repotting. Plant largest out and grow smallest on 5 cm deep, 5 cm apart, in boxes.	Late September/October	Plant outdoors 10 cm deep, 10 cm apart, in mild areas, or 5–6 corms to a 13 cm pot in frost-free greenhouse elsewhere. Small corms will flower in 2 years, seedlings in 3–4. Grow seedlings on as for offsets after first season's growth.
	SEED	Sow in slight heat and topdress with a little compost in autumn.	April	
Tropaeolum (tuberous-rooted types; T. speciosum = Chilean Flame Flower)	DIVISION (of tuber)	Lift and divide older plants replanting immediately 10 cm deep, 30 cm apart, in a northerly aspect.	In autumn or early spring	Climbers for trellis or through hedges etc. Require a cool root run in well-drained but leafy type soil. Pot seedlings on carefully to avoid root disturbance and overwinter in cold frame.
	SEED	Sow singly in small pots in slight heat.	June	
Tulipa (Tulips)	DIVISION (offsets; bulbs)	Detach offsets on lifting and replant in autumn 5 cm deep, 8 cm apart, in boxes or prepared beds in sun outdoors. Lift during summer each year and dry off.	As foliage dies back after flowering	Ideal bulbs for bedding, containers or as cut flower. Feed offsets during growing period and remove any flowers as they form. Bulbs will be ready for normal treatment in 2 years.
Urceolina	DIVISION (offsets; bulbs)	Remove offsets when repotting and grow on singly in small pots. Will flower in 2–3 years.	September/October	Requires only frost-free temperatures under glass planting 3 bulbs to a 13 cm pot.

PLANT NAME	METHOD OF PROPAGATION	TREATMENT REQUIRED	TIME OF YEAR PROPAGATED	CULTURAL NOTES
Urginea *(U. maritima = Sea Onion)*	DIVISION (offsets; bulbs)	Lift established groups and remove offsets. Replant 15 cm deep, 20 cm apart. Offsets will flower 1–2 years from separation.	In autumn	Only suitable for gardens in the mildest areas. Handle bulbs carefully as sap is a skin irritant. Protect bulbs from frost during winter.
Vallota *(Scarborough Lily)*	DIVISION (offsets) bulbs)	Shake older plants from pot and carefully remove offsets. Grow on in small pots. Rest during May and June when pots are best placed outdoors in sun.	Mid-summer before flowering	Most suitable as a houseplant in a semi-shaded position. Pot on as necessary.
Veltheimia	DIVISION (offsets; bulb)	Remove offsets when repotting and grow on in small pots.	Early in autumn	Grow in a cool greenhouse just keeping frost at bay or in a sunny window as a houseplant. Offsets will flower in four years or so. Root leaves in a shaded propagator and maintain a humid atmosphere.
	LEAF CUTTINGS (produce bulbils)	Remove mature leaves and plant upright in pots. Bulbils will form at the base in a month or so. Grow on as for offsets.	In spring or summer	
Watsonia *(Southern Bugle Lily)*	DIVISION (offsets; corms)	Detach cormlets when lifting or repotting and grow on in boxes under glass. Divide evergreens every 4th year in spring.	Late autumn	Deciduous species are fairly hardy; evergreen types are tender. Outside plant 10 cm deep, 10 cm apart, or more if growing on *in situ*. Under glass plant 3–4 to a 13 cm pot. After 2 years pot seedling cormlets singly in small pots to grow on.
	SEED	Sow in slight heat and grow on undisturbed for 2 years. Topdress annually as leaves die back.	In spring	
Zantedeschia *(Arum Lily; Z. aethiopica, syn. Richardia aethiopica)*	SEED (soak 24 hours in tepid water before sowing)	Sow in propagator at 21°C and grow on in small pots.	In spring or early summer	Pot plants for the warm greenhouse in all but very mildest areas. Plant 1 to a 15–18 cm pot.
	DIVISION (of rhizome)	Divide when repotting.	August/early September	
Zephyranthes *(Flowers of the Western Wind)*	DIVISION (offsets; bulbs)	Lift or repot older bulbs replanting larger offsets 4–5 to a 13 cm pot to grow on to flowering. Smaller sizes are best grown on in boxes for a couple of years.	After flowering	Plants with crocus-like flowers best grown in a cold greenhouse in all but the very mildest areas where they can be planted out 10 cm deep, 15 cm apart. Disturb as little as possible and do not dry off between plantings.
Zygadenus	DIVISION (offsets; bulbs)	Lift established groups, remove offsets and replant 8 cm deep, 23 cm apart.	In autumn	Plants for open woodland conditions of semi-shade and leafy type soil. Offsets will flower in 2 years.

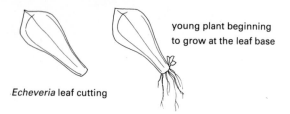

Echeveria leaf cutting

young plant beginning
to grow at the leaf base

A type of wedge graft:
useful for grafting
pendulous cacti or
succulents onto an
upright stock to create
a 'standard' plant

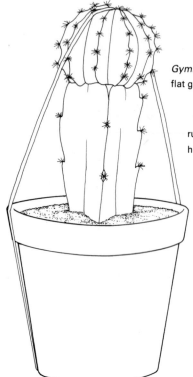

Gymnocalycium mihanovichii
flat grafted onto *Hylocereus*

rubber band
holding graft in place

young plantlets fall from
leaf margins and form roots
when they come to rest
on the compost surface

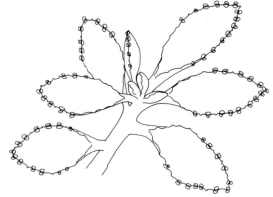

Plantlets developing on the leaf margins
of a *Kalanchoe* (*Bryophyllum*)

Roots forming on a
'dry' cutting offset
of *Chamaecereus*
prior to planting

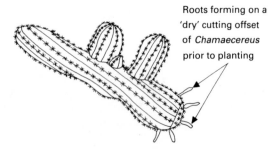

11 Cacti and Succulents

The majority of cacti and succulents rank among the easiest of all plants to propagate, indeed several succulents produce plantlets on their leaves which fall to the ground and root of their own accord. It is also true to say that almost all cacti and succulents can be raised from seed, but in practice many germinate and/or grow so slowly that it is more practical to propagate them vegetatively. For sowing seed, John Innes seed compost with the addition of a little coarse sand to ensure perfect drainage is suitable. Small seed should be surface-sown, whereas larger seeds can be pressed into the compost surface slightly. A propagator heated to 18–24 °C is necessary for good germination, and the seedlings are best grown on undisturbed for their first year before being transplanted or potted on. Be guided by their rate of growth. 'Damping off' is likely to be their greatest enemy, so watch out for this. Sowing thinly and keeping the compost moist, but never overwet, will make losses less likely. Root cuttings in a similar position and shade both these and seedlings from hot sunshine until well established.

Potting compost requirements for these plants are so diverse that I can only give a general guide here, and advise once more the use of a John Innes potting compost with the addition of a quantity of coarse sand. Use compost No. 1 for desert cacti and Nos. 2 or 3 for succulents and epiphytes, which usually require a compost richer in food and water-retentive properties.

During winter most cacti and succulents require a rest period when they need little or no water and very little heat. Unless otherwise stated, the plants in the following list are best grown in a greenhouse with a minimum winter temperature of 7 °C, and a sunny position during summer.

A brief note on grafting may be useful as this is used on a number of cacti for a variety of reasons, the main being to speed on the growth of slow-growing species or to create a 'standard' with a naturally pendulous species grafted onto an upright stock.

The same principles of cambial contact and cleanliness apply to grafting cacti as with other plants and it is necessary to work fairly quickly to avoid the cut surfaces drying out before contact is made.

A flat graft is most often used for the former reason, and plants can be grafted from seedlings one year old onwards, ideally using rootstocks and scions of the same diameter. They are simply cut across straight at the required spot with a razorblade and fitted together. Hold in place either with a couple of rubber bands or a cactus spine pushed through the side of the scion, the former being preferable, in my opinion, as it causes no damage if done correctly. Use a small cloth pad on top of the scion if necessary.

More slender species used as 'standards' are best wedge grafted, perhaps two or three scions being placed around the rootstock to make a 'head'. The scions in this case are held in place with a cactus spine. Keep all grafts in a warm dry atmosphere until a union has been made. This usually occurs within the space of a few weeks or so.

PLANT NAME	METHOD OF PROPAGATION	TREATMENT REQUIRED	TIME OF YEAR PROPAGATED	CULTURAL NOTES
Adenium	HALF-RIPE CUTTINGS	Root in a warm propagator.	Spring or early summer	Medium-sized succulent shrubs for the greenhouse.
Adromischus	LEAF CUTTINGS	Push leaf slightly into the compost surface at an oblique angle. Plant 2.5 cm apart.	During summer	Small slow-growing succulents making good houseplants. Require a winter minimum temperature of 10°C.
Aeonium	STEM CUTTINGS	Cut stem below a rosette, leave cut to dry for 24 hours then plant as a cutting.	In spring or summer	Succulents looking like tall sempervivums. Quite hardy. Suitable for cold greenhouses or as houseplant.
	SEED	Sow in a temperature of 18–21°C.	In spring	
Agave (American Aloe/Century Plant)	DIVISION (offsets)	Detach offsets and grow on in small pots.	During spring or summer	Rosette-forming succulents, some growing very large. Place outdoors in summer but keep frost-free during winter.
	SEED	Sow in warm propagator.	In spring	
Aichryson	SEED	Sow in a temperature of 18–21°C.	February to May	Rosette-forming succulents similar to *Aeonium* in looks and culture.
Aloe (A. variegata = Partridge Breasted Aloe)	DIVISION (offsets)	Root in a warm shady propagator or pot singly if offsets rooted when removed.	In spring or early summer.	Upright rosette-forming succulents making ideal houseplants in slightly shaded positions.
Anacampseros	SEED	Sow in a temperature of 18–21°C.	As soon as ripe or in spring	Small slow-growing succulents requiring a compost rich in lime.
Ancistrocactus	SEED	Sow in a temperature of 21–24°C.	February to May	Cylindrical cacti once included with *Echinocactus* and needing the same culture.
Aporocactus (A. flagelliformis = Rat Tailed Cactus)	SOFTWOOD CUTTINGS	Cut young shoots into short sections. Dry in warm spot for 48 hours before placing on compost surface to root.	During summer	Hanging cacti ideal for hanging baskets or shelves in house or greenhouse. Water frequently when growing. To grow as standard graft onto 20 cm stems of *Nyctocereus serpentinus*.
	SEED	Sow in a temperature of 21–24°C.	February to May	
Argyroderma	SEED	Sow in a temperature of 21–24°C and keep atmosphere dry when seedlings appear.	February to May	Pebble-like succulents for full sun and a winter minimum of 10°C. Keep dry December to April.
Ariocarpus	SEED	Sow in a temperature of 21–24°C. Seedlings grow very slowly.	February to May	Spineless cacti needing deep pots to accommodate long roots.
Astroloba	DIVISION (offsets)	Remove offsets when available and pot on in small pots.	During summer	Rosette-forming succulents similar to *Aloe*. Water freely in summer little in winter.
Astrophytum (Star Cactus/Bishop's Cap)	SEED	Sow in a temperature of 21–24°C.	February to May	Easily grown houseplant cacti. Require a compost rich in lime.
Blossfeldia	SEED (slow germination)	Sow in a temperature of 21–24°C.	February to May	Minute cacti among the smallest of flowering plants and very slow-growing.
Borzicactus	SEED	Sow in a temperature of 21–24°C.	February to May	Low-growing cylindrical cacti. Ideal houseplant for sun or slight shade.
Cactus (Turk's Cap Cactus; includes Melocactus)	SEED	Sow in a temperature of 21–24°C.	February to May	Mainly difficult and slow-growing cacti with few roots. Require warm winter temperature of 20°C.
	FLAT GRAFTING	Graft seedlings onto *Trichocereus spachianus* for faster growth.	May/June	

PLANT NAME	METHOD OF PROPAGATION	TREATMENT REQUIRED	TIME OF YEAR PROPAGATED	CULTURAL NOTES
Caralluma	SEED	Sow in a temperature of 21–24°C.	February to May	Almost leafless succulents. Grow in slight shade in summer and avoid overwatering. Flowers are similar to *Stapelia*.
	SOFTWOOD CUTTINGS	Make cuttings from young shoots. Dry 48 hours then place on compost surface to root.	May to September	
Carnegiea (C. gigantea, syn. Cereus gigantea)	SEED	Sow in a temperature of 18–21°C.	February to May	Tall, slow-growing columnar cacti for a warm spot and a 13°C winter minimum.
Carpobrotus	SEED	Sow in a temperature of 18–21°C.	February to May	Low-growing succulent with trailing stems. Grow outside in summer frost-free spot in winter.
	TIP CUTTINGS	Root small cuttings in a warm propagator.	During summer	
Cephalocereus	SEED	Sow in a temperature of 21–24°C.	February to May	Columnar cacti with whitish hairs. Only requires frost-free winter heat. Keep dry until spring.
	SOFTWOOD CUTTINGS	Make cuttings from tops or side branches. Dry for 48 hours then plant upright.	May to September	
Cephalophyllum	SEED	Sow in a temperature of 21–24°C.	February to May	Low-growing rosette-forming succulents for house or greenhouse.
Cereus (C. pentagonus, syn. Acanthocereus pentagonus)	SEED	Sow in a temperature of 21–24°C.	February to May	Columnar cacti making good houseplants. *C. jamacaru* and *C. peruvianus* are often used as rootstocks.
Ceropegia (C. woodii = Hearts Entangled)	SOFTWOOD CUTTINGS	Use small side shoots and root in warm propagator.	April/May	Trailing or climbing succulents. Water well in summer little in winter. *C. woodii* bears corms at some branch endings and may be propagated from these.
	SEED	Sow in a temperature of 21–24°C.	February to May	
Chamaecereus (C. silvestrii = Peanut Cactus)	SOFTWOOD CUTTINGS	Remove short sections of stem at a joint. Leave 48 hours to dry then place on compost to root in warm place.	During summer	Low-growing much-branched cacti. Ideal houseplant. Very easy to manage.
Cheiridopsis	SEED	Sow in a temperature of 21–24°C.	February to May	Small succulent houseplants requiring very little water.
Cleistocactus (C. strausii = Silver Torch)	SEED	Sow in a temperature of 21–24°C.	February to May	Columnar cacti ideal as houseplants. Need only frost-free conditions during winter.
	DIVISION (offsets)	Detach offsets and treat as cuttings.	During spring and summer.	
Conophytum (includes Ophthalmophyllum)	SEED	Sow at 21–24°C and grow on in very small pots.	February to May	Low growing succulents similar to *Lithops*. Keep dry in May and June cool but frost-free in winter. Houseplants.
Copiapoa	SEED	Sow in a temperature of 21–24°C.	February to May	Globular cacti growing best as grafted plants.
	FLAT GRAFTING	Graft onto a *Cereus* rootstock at the seedling stage.	May/June	
Corryocactus	SEED	Sow in a temperature of 21–24°C.	February to May	Rather tall branching columnar cacti for a large greenhouse.
Coryphantha	SEED	Sow in a temperature of 21–24°C.	February to May	Slow-growing globular or cylindrical cacti. Keep cool but frost-free in winter.
Cotyledon	SEED (seedlings may be variable)	Sow in a temperature of 21–24°C.	February to May	Most in cultivation are shrubby succulents. Add old mortar rubble to the compost and keep dry in winter.
	LEAF CUTTINGS	Use leaves or rosette of leaves at the top of the stem to make cuttings.	June to September	

Cacti and Succulents 97

PLANT NAME	METHOD OF PROPAGATION	TREATMENT REQUIRED	TIME OF YEAR PROPAGATED	CULTURAL NOTES
Crassula	SEED	Sow at 21–24°C. Seedlings may be variable.	February to May	Much branched and often shrubby succulents. Ideal houseplants. Give plenty of water in summer and a 10°C minimum in winter.
	SOFTWOOD CUTTINGS	Use leaves with stalk or short sections of stem as cuttings.	Late spring or summer	
Delosperma	SEED	Sow in a temperature of 21–24°C.	February to May	Dwarf succulent shrubs which can be planted out in the rockery during summer.
	TIP CUTTINGS	Remove short non-flowering shoots and root in slight warmth.	During summer	
Dinteranthus	SEED	Sow in a temperature of 21–24°C.	February to May	Pebble-like succulents. Keep dry and frost-free in winter. Water well in summer.
Disocactus	SEED	Sow in a temperature of 21–24°C.	February to May	Epiphytic cacti which flower when quite small. Keep warm in winter.
Dolichothele	SEED	Sow in a temperature of 21–24°C.	February to May	Barrel or globular shaped cacti best grown in a greenhouse in slight shade.
Drosanthemum	SEED	Sow in a temperature of 18–21°C.	February to May	Shrubby succulents for outdoors or a cold greenhouse. Can be used as bedding plants.
	SOFTWOOD CUTTINGS	Root short pieces in a peat/sand compost and spray with water occasionally until rooted.	May to September	
Duvalia	SOFTWOOD CUTTINGS	Use short sections of new growth. Dry for 48 hours then place on surface of compost of peat/sand to root.	May to September	Leafless succulents similar to *Stapelia*. Make reasonable houseplants. Slight shade in summer.
Eccremocactus	SEED	Sow in a temperature of 21–24°C.	February to May	Epiphytic cacti with hanging stems most suited to a warm greenhouse.
Echeveria (includes Dudleya)	DIVISION (offsets)	Remove rosette of the offsets and treat as cuttings.	May to September	Low-growing rosette-forming succulents. Avoid overwatering at any time. Keep cool but frost-free in winter. *E. glauca* can be used as a summer bedding plant.
	LEAF CUTTINGS	Detach leaves and dib into a peat/sand mixture to root.	May to September	
Echidnopsis	SOFTWOOD CUTTINGS	Use short section of stem drying for 48 hours before placing on the surface of compost to root.	May to September	Leafless succulents. Grow in slight shade in summer and water only occasionally.
Echinocactus (Golden Barrel)	SEED	Sow in a temperature of 21–24°C.	February to May	Slow-growing barrel shaped cacti. Ideal for the house and a 10°C minimum in winter.
Echinocereus (E. rigidissimus = Rainbow Cactus; E. triglochidiatus = Strawberry Cactus)	SEED	Sow in a temperature of 21–24°C.	February to May	Mostly easily grown clump-forming cacti. Grow as houseplants or in greenhouse cool dry and frost-free in winter.
	DIVISION (suckers)	Remove suckers when available and pot singly.	Spring through summer	
	SOFTWOOD CUTTINGS	Separate and use short stems to form cuttings.	May to September	
Echinofossulo-cactus (Stenocactus)	SEED	Sow in a temperature of 21–24°C.	February to May	Spherical shaped cacti for house or greenhouse. Grow in slight shade in summer.
Echinopsis (includes Trichocereus which are taller types; E. violacea, syn. Acanthocalycium violaceum)	DIVISION (offsets)	Remove and replant offsets in small pots to grow on.	Spring through summer	Globular cacti requiring a minimum of heat in winter if kept dry. Do not overpot. *Trichocereus spachianus* is often used as a rootstock.
	SEED	Sow in a temperature of 21–24°C.	February to May	

PLANT NAME	METHOD OF PROPAGATION	TREATMENT REQUIRED	TIME OF YEAR PROPAGATED	CULTURAL NOTES
Epiphyllum (Orchid Cactus; includes Phyllocactus)	SEED	Sow in a temperature of 21–24°C.	February to May	Epiphytic cacti making excellent houseplants. Grow outside in a shady spot in summer and house in a warm light position in the winter.
	SOFTWOOD CUTTINGS	Take cuttings across a broad section at the top of a stem and root in a peat/sand mixture in slight heat.	May to September	
Escobaria	SEED	Sow in a temperature of 21–24°C.	February to May	Globular or cylindrical cluster-forming cacti.
	DIVISION (offsets)	Remove offsets and treat as cuttings.	May to September	
Escontria	SEED	Sow in a temperature of 21–24°C.	February to May	Tree-like branching cacti with edible fruit. Best grown from seed.
Espostoa	SEED	Sow in a temperature of 21–24°C.	February to May	Columnar cacti with silky hairs. Grow in sun or slight shade.
	FLAT GRAFTING	E. lanata can be grafted onto Trichocereus spachianus.	May/June	
Euphorbia (E. millii = Crown of Thorns)	SEED (wash largest seed in clean water before sowing)	Sow in a temperature of 21–24°C.	February to May	A varied genus with many forms suitable as houseplants. Sap is a contact irritant so avoid getting it on the skin or in eyes. Plants need cool conditions in winter in most cases.
	STEM CUTTINGS	Dip cuttings in powdered charcoal to stop bleeding. Root in sand/peat compost.	May to September	
Facheiroa	SEED	Sow in a temperature of 21–24°C.	February to May	Tree-like much-branched cacti.
Faucaria (Tiger's Chaps)	SEED	Sow in a temperature of 21–24°C.	February to May	Low-growing succulents making good houseplants. Water well in summer, dry out in winter.
	DIVISION (offsets)	Remove offsets when repotting and pot on singly.	March/April	
Fenestraria (Window Plants)	SEED	Sow in a temperature of 21–24°C.	February to May	Extreme pebble-like succulents with transparent leaf tips. Require a minimum temperature of 16°C in winter.
	DIVISION	Divide older groups and repot carefully.	April when repotting	
Ferocactus	SEED	Sow in a temperature of 21–24°C.	February to May	Slow-growing barrel-shaped cacti. Very spiny but easily grown.
Gasteria	DIVISION (offsets)	Remove offsets and pot singly. They are usually rooted when separated.	Late spring and summer	Rosette-forming succulents making good houseplants, especially for narrow windowsills. Only need frost-free conditions in winter.
	SEED	Sow in a temperature of 21–24°C.	February to May	
Gibbaeum	SEED	Sow in a temperature of 21–24°C.	February to May	Small succulents which have definite rest periods which vary with the species. Require a 16°C minimum in winter.
	DIVISION (offsets)	Detach offsets and dry for 48 hours before planting as cuttings.	March when repotting	
Glottiphyllum	SEED	Sow at 21–24°C. Seedlings may be variable.	February to May	Low-growing succulents requiring extremely poor soil and very little water. Can be used as houseplants.
	TIP CUTTINGS	Cut short stems to form cuttings or plants can be divided.	During summer	
Greenovia	SEED	Sow in a temperature of 21–24°C.	February to May	Rosette-forming succulents making good houseplants. Keep cool but frost-free in winter.
	STEM CUTTING	Use entire rosette as a cutting and root in slight heat.	May to September	
Grusonia	SOFTWOOD CUTTINGS	Make cuttings of two sections of stem cutting at joint. Root in peat and sand.	May to September	Shrubby cacti with cylindrical stems.

PLANT NAME	METHOD OF PROPAGATION	TREATMENT REQUIRED	TIME OF YEAR PROPAGATED	CULTURAL NOTES
Gymnocalycium	SEED	Sow in a temperature of 18–21°C.	February to May	Spherical cacti making good houseplants. Give plenty of water in growing season.
	FLAT GRAFTING	*G. mihanovitchi* is often grafted onto *Hylocereus quatemalense*.	May/June	
Haageocereus	SEED	Sow in a temperature of 21–24°C.	February to May	Easily-grown columnar cacti often with coloured spines. Grafting makes for more even growth.
	FLAT GRAFTING	Graft onto *Eriocereus jusbertii*.	May/June	
Hamatocactus	SEED	Sow in a temperature of 21–24°C.	February to May	Globular cacti needing only frost-free conditions in winter when they are kept dry.
Hariota (syn. *Hatiora*)	SEED	Sow in a temperature of 21–24°C. Keep compost moist at all times.	February to May	Epiphytic cacti making ideal hanging basket or houseplant for semi-shaded position.
Harrisia (syn. *Eriocereus*)	SEED	Sow in a temperature of 21–24°C.	February to May	Long-stemmed prostrate cacti with nocturnal flowers. Keep cool but frost-free in winter, warm in summer.
	STEM CUTTINGS	Use sideshoots or sections of stem and dry for up to a week before placing in sand to root.	May to September	
Haworthia	SEED	Sow at 21–24°C. Seedlings may be variable.	February to May	Rosette-forming succulents ideal in the house in sun or slight shade. Grow in pans or half-pots.
	DIVISION (offsets)	Detach and grow on in small pots. Usually rooted on separation.	Late spring or early summer	
Heliocereus (Sun Cactus)	SEED	Sow in a temperature of 21–24°C.	February to May	Trailing cacti for semi-shade. Make good houseplants and need a 10°C minimum temperature in winter.
	FLAT GRAFTING	Graft onto *Cereus hassleri* to grow as a standard.	May/June	
Hereroa	SEED	Sow in a temperature of 21–24°C.	February to May	Low-growing succulent plants for the house. Keep dry in winter.
Homalocephala	SEED	Sow at 21–24°C. Seedlings slow to grow at first.	February to May	Easy free-flowering globular cacti. Grow in the house in full sun.
Huernia (Carrion Flowers)	SEED	Sow in a temperature of 18–21°C.	February to May	Low-growing succulents with flowers smelling of rotten meat. Avoid overwatering and give a minimum heat of 10°C in winter.
	STEM CUTTINGS	Use small sections of new growth dried for 48 hours before being placed on compost surface to root.	May to September	
Hylocereus (Wood Cereus)	SEED	Sow at 18–21°C.	February to May	Epiphytic climbing cacti for semi-shade in a warm moist atmosphere.
	STEM CUTTINGS	Cut stems at joints and make cuttings of 1 or 2 sections.	May to September	
Jatropha (*J. podagrica* = Guatemala Rhubarb)	SEED	Sow at a temperature of 21–24°C.	February to May	Shrubby succulent plants related to *Euphorbia*. House or greenhouse plants.
Kalanchoe (includes *Bryophyllum* and *Kitchingia*)	SEED	Sow in a temperature of 18–21°C.	February to May	Ideal succulent houseplants for a 10°C winter minimum. Some species produce plantlets on their leaves which can be removed and grown on.
	SOFTWOOD CUTTINGS	Make cuttings 5 cm long from non-flowering shoots. Dry 48 hours before dibbing in compost to root.	July/August	
Kleinia (*K. articulata* = Candle Plant)	SOFTWOOD CUTTINGS	Detach shoots at joints or cut sections from stem and dry 24 hours before placing in compost to root.	April to June	Succulents ideal as houseplants. Keep cool but frost-free in winter.

PLANT NAME	METHOD OF PROPAGATION	TREATMENT REQUIRED	TIME OF YEAR PROPAGATED	CULTURAL NOTES
Lampranthus (L. roseum = Midday Flower)	SEED SOFTWOOD CUTTINGS	Sow in a temperature of 21–24°C. Use shoots 5 cm long and root and grow on several to a pot.	February to May Spring and summer	Low-growing shrubby succulents. Can be used for summer bedding and housed in a frost-free place in winter.
Lemaireocereus (includes Stenocereus. L. marginatus = Organ Pipe Cactus)	SEED STEM CUTTINGS	Sow in a temperature of 21–24°C. Use sideshoots and root in sand drying for 48 hours before planting.	February to May May to September	Tree-like clump-forming cacti with colourful spines. Grow in warm sunny positions with winter minimum of 13°C.
Leuchtenbergia	SEED	Sow in a temperature of 18–21°C. Seedlings will flower in 4–5 years.	February to May	Unusual cacti rather like an Aloe. Grow in a deep pot in a warm place. Keep fairly dry.
Lithops (Living Stones/Pebble Plants)	SEED (surface sow) DIVISION	Sow in a temperature of 18–21°C. Divide and repot older clumps.	February to May May/June	Pebble-like succulents which make good houseplants. Keep warm in summer, cool but frost-free in winter. Do not overwater.
Lobivia	SEED DIVISION (offsets)	Sow in a temperature of 21–24°C. Remove offsets and treat as cuttings.	February to May May to August	Mostly cylindrical cacti for the house or garden where it only needs protection from frost and heavy rain.
Lophocereus	SEED STEM CUTTINGS	Sow in a temperature of 21–24°C. Use sideshoots as cuttings and root in a sand/peat compost.	February to May May to September	Tall columnar cacti easy to cultivate in sun or slight shade.
Lophophora (L. williamsii = Dumpling Cactus/Mescal Button)	SEED	Sow at 21–24°C. Seedlings are very slow-growing.	February to May	Low-growing round cacti. Grow in deep pots. Keep dry and frost-free in winter.
Maihuenia	STEM CUTTINGS	Make cuttings from lengths of stem with at least 1 joint. Root in peat and sand.	May to September	Low-branched and bushy cacti with persistent leaves. Suitable for a cold house in mild areas.
Malacocarpus	SEED	Sow in a temperature of 21–24°C.	February to May	Cacti with globular or cylindrical stems. Easy to grow and quite hardy if kept dry in winter.
Malephora	SEED	Sow in a temperature of 21–24°C.	February to May	Shrubby succulent suitable for cold greenhouse.
Mammillaria (Pincushion. M. plumosa = Feather Cactus)	DIVISION (offsets) SEED	Remove offsets at their base and treat as cuttings rooting in a peat/sand compost. Sow at 21–24°C. Seedlings often very slow growing.	May to September February to May	Many species of globular or cylindrical cacti making ideal houseplants. Water sparingly.
Mammillopsis	SEED DIVISION (offsets)	Sow in a temperature of 21–24°C. Remove offsets and root in sand. Pot on singly in small pots.	February to May May to September	Cluster-forming globular or cylindrical cacti. Keep cool but frost-free during winter when they should be dry.
Mila	SEED	Sow at a temperature of 21–24°C.	February to May	Small clump-forming cacti with cylindrical stems. Easy to grow.
Monanthes	SEED LEAF or STEM CUTTINGS	Sow in a temperature of 21–24°C. Use leaves or short sections of stem and root in peat/sand.	February to March May to September	Dwarf-branching or creeping succulents for house or greenhouse. Grow in slight shade in summer and a little warmth in winter.

PLANT NAME	METHOD OF PROPAGATION	TREATMENT REQUIRED	TIME OF YEAR PROPAGATED	CULTURAL NOTES
Monvillea	SEED	Sow in a temperature of 21–24°C.	February to May	Slender-stemmed prostrate cacti best suited to a warm greenhouse.
	STEM CUTTINGS	Make cuttings from sections of stem and dry for up to 1 week. Keep dry when planted until roots form.	May to September	
Myrtillocactus	SEED	Sow in a temperature of 21–24°C.	February to May	Much-branched columnar cacti for semi-shade in a warm greenhouse. Keep dry in winter.
	STEM CUTTINGS	Use sideshoots as stem cuttings cutting at the joint and drying out before planting in peat/sand.	May to September	
Nananthus	SEED	Sow in a temperature of 21–24°C.	February to May	Minute succulents with long roots needing long pots in which to grow. Keep warm in winter.
Neohenrica	SEED	Sow in a temperature of 21–24°C.	February to May	Mat-forming succulents for the greenhouse. Avoid watering the rosettes.
	DIVISION	Divide mats and repot separately.	May to September	
Neolloydia	SEED	Sow in a temperature of 21–24°C.	February to May	Small cylindrical cacti usually forming clusters. Keep dry in winter.
	STEM CUTTINGS	Use sideshoots as cuttings and root in a mixture of peat/sand.	May to September	
Neoporteria	SEED	Sow in a temperature of 21–24°C.	February to May	Mostly cylindrical cacti many of which grow better as houseplants if grafted. Grow in large pot if on its own roots.
	FLAT GRAFTING	Graft onto plants of *Eriocereus jusbertii*.	May/June	
Nopalxochia	SOFTWOOD CUTTINGS	Take cuttings from top of the stem and dry slightly before rooting in peat/sand. Spray over occasionally until rooted.	May to September	Epiphytic cacti similar to *Epiphyllum*. Grow in a warm slightly shaded spot and water fairly frequently. Can also be raised from seed.
	WEDGE GRAFTING	Graft onto *Hylocereus undatus* for greater growth rate.	May/June	
Notocactus	SEED	Sow in a temperature of 21–24°C.	February to May	Easily-grown barrel shaped cacti making good houseplants. Do not overwater and keep warm in winter.
	FLAT GRAFTING	Graft cristate forms onto *Cereus peruvianus*.	May/June	
Nyctocereus	SEED	Sow in a temperature of 21–24°C.	February to May	Slender-stemmed cacti which are upright when young and trailing later.
	SOFTWOOD CUTTINGS	Use sideshoots as cuttings drying for up to a week before placing on sand to root.	May to September	
Obregonia	SEED	Sow in a temperature of 21–24°C.	February to May	Cacti with a flat globose body and a large tap root. Grow in deep pots.
Opuntia (O. ficus-indica = Prickly Pear; O. microdasys = Rabbit's Ears)	SEED (soak 48 hours in tepid water before sowing)	Sow in a temperature of 21–24°C.	February to May	Large genus of many types. Most common are those with flat oval-jointed stems. Some species such as *O. polycantha* are hardy in dry mild areas.
	STEM CUTTINGS	Use mature 'pads' separated from the parent plant, dried for 48 hours, then rooted in sand.	May to September	
Oreocereus (O. trollii = Old Man of the Andes)	SEED	Sow in a temperature of 18–21°C.	February to May	Columnar cacti making good houseplants. Keep cool and dry in winter and add limestone chippings to compost.
Oroya	SEED	Sow in a temperature of 21–24°C.	February to May	Low-growing globular cacti for the cold greenhouse.

PLANT NAME	METHOD OF PROPAGATION	TREATMENT REQUIRED	TIME OF YEAR PROPAGATED	CULTURAL NOTES
Oscularia	TIP CUTTINGS	Root in a sand/peat compost in slight heat.	During summer	Shrubby creeping or hanging succulents. Grow outdoors in summer, dry and frost-free in winter.
Pachycereus	SEED	Sow in a temperature of 21–24°C.	February to May	Tall branched tree-like cacti which grow very slowly.
Pachyphytum (P. oviferum = Sugar Almond Plant)	LEAF CUTTINGS	Detach leaves and insert into peat/sand compost at a slight angle or use entire rosette as a cutting.	May to September	Shrubby succulents forming rosettes of chubby leaves. Only needs frost-free winter temperature.
Pachypodium	SEED	Sow in a temperature of 21–24°C.	February to May	Rare shrubby succulents with a trunk-like stem.
Parodia (includes Islaya)	SEED	Sow at 18–21°C. Many seedlings will begin to flower when 3–4 years old.	February to May	Small slow-growing barrel shaped cacti requiring a minimum of 13°C in winter.
Pectinaria	STEM CUTTINGS	Root short cuttings in a sand/peat compost in a warm propagator.	May to September	Low growing leafless succulents growing best in a warm spot in semi-shade.
Pediocactus	SEED	Sow in a temperature of 21–24°C.	February to May	Globular cacti similar to *Mammillaria* and sometimes forming into groups.
Pelecyphora	SEED	Sow at 21–24°C. Seedlings are very slow-growing.	February to May	Slow-growing club shaped cacti-forming clusters. Do not overwater.
Pereskia (P. aculeata = Barbados Gooseberry)	TIP CUTTINGS	Root in a peat/sand compost in a warm propagator.	May to September	Shrubby cacti with evergreen leaves requiring a minimum winter temperature of 13°C. Mostly used as a rootstock.
	SEED	Sow in a temperature of 21–24°C.	As soon as ripe	
Pfeiffera	SEED	Sow at 21–24°C. Seedlings can be wedge grafted to *Cereus jamacaru* if required.	February to May	Small epiphytic hanging cacti best suited to a warm greenhouse in slight shade.
Pleiospilos	SEED	Sow at 18–21°C. Pot seedlings in deep pots to accommodate roots.	February to May	Succulent mimicry plants similar to *Lithops*. Water freely May/June, little otherwise. Make good houseplants.
Porfiria	SEED	Sow in a temperature of 21–24°C.	February to May	Round cacti rarely producing offsets. Best grown in a warm greenhouse.
Portulacaria	STEM CUTTINGS	Use small sideshoots and dry for 24 hours before planting in a peat/sand compost.	May to September	Shrubby tree-like succulents. Avoid overwatering and keep dry in winter.
Pterocactus	SEED	Sow at 21–24°C. Seedlings can be grafted onto *Opuntia* if so wished.	February to May	Bushy cactus branching from the ground. Do not overwater.
Quiabentia	SEED	Sow in a temperature of 21–24°C.	February to May	Straggly tree-like or bushy cacti best suited to a large greenhouse.
Rathbunia	SEED	Sow in a temperature of 21–24°C.	February to May	Cacti eventually forming into fairly large straggly bushes.
Rebutia	SEED	Sow at 18–21°C. Seedlings are better than offsets.	February to May	Barrel or globular cacti making good houseplants and only requiring frost-free conditions during the winter.
	DIVISION (offsets)	Remove offsets being careful not to damage the parent and treat as cuttings.	May to September	

PLANT NAME	METHOD OF PROPAGATION	TREATMENT REQUIRED	TIME OF YEAR PROPAGATED	CULTURAL NOTES
Rhipsalidopsis (including Easter Cactus)	STEM CUTTINGS	Make cuttings of stems to include at least one joint, preferably two. Root in a peat/sand compost.	May to August	Epiphytic cacti best grown in semi-shade outdoors through summer and in a warm room during winter. Make ideal houseplants.
Rhipsalis (includes Lepismium)	SEED	Sow in a temperature of 21–24°C.	February to May	Epiphytic cacti which need a rich compost and a warm moist atmosphere.
	STEM CUTTINGS	Root lengths of stem containing at least 1 segment in peat/sand.	May to September	
Rochea	SEED	Sow at 21–24°C. Inclined to be variable.	February to May	Shrubby succulents making good houseplants. Water sparingly but regularly during the year. Do not water cuttings until roots have formed.
	TIP CUTTINGS	Root several to a pot and grow on as a group.	May/June	
Ruschia	SEED	Sow in a temperature of 21–24°C.	February to May	Low-growing succulents for house or greenhouse. Keep dry from January to June and frost-free in winter.
Sarcocaulon	ROOT CUTTINGS	Root in a sandy compost in a warm propagator.	February/March	Spiny sub-shrubs requiring the same kind of cultivation as cacti.
Sarcostemma	SEED	Sow in a temperature of 21–24°C.	February to May	Climbing or straggling succulents for a warm greenhouse.
Schlumbergia (Christmas cactus; includes Zygocactus)	STEM CUTTINGS	Make cuttings of stems to include at least one joint, preferably two. Root in a peat/sand compost.	May to August	Epiphytic cacti best grown in semi-shade outdoors through summer and in a warm room during winter. Make ideal houseplants.
	CLEFT GRAFTING	Graft shoots onto *Pereskia aculeata* to create a 'standard'	May/June	
Selenicereus (S. grandiflorus = Queen of the Night)	SEED	Sow in a temperature of 21–24°C.	February to May	Large climbing cacti with slender stems. Require warm greenhouse cultivation. Flowers open at night.
	STEM CUTTINGS	Use sections from young shoots and root on peat/sand. Spray over with water occasionally until rooted.	May to September	
Stapelia (Carrion Flower)	SEED (germinates very quickly)	Sow in a temperature of 18–21°C.	February to May	Low-growing succulents with splendid flowers but a smell of rotten meat. Make good houseplants in airy conditions and a 10°C minimum temperature.
	STEM CUTTINGS	Use small sections of new growth, dry for 48 hours then place on a compost of peat/sand to root.	May to September	
Stetsonia	SEED	Sow in a temperature of 21–24°C.	February to May	Tree-like cacti. Keep dry in winter and fairly warm.
	FLAT GRAFTING	Seedlings can be grafted onto *Eriocereus jusbertii*.	May/June	
Strombocactus	SEED	Sow in a temperature of 21–24°C.	February to May	Small flat cacti which prefer a compost rich in lime and a warm position all year.
Synadenium	STEM CUTTINGS	Use sections of young shoots and dry for 24 hours before inserting in sand to root.	May to September	Succulent shrubs related to *Euphorbia* and requiring similar cultivation.
Tavaresia	SEED	Sow in a temperature of 21–24°C.	February to May	Leafless succulents making good houseplants. Grow as for *Stapelia*.
Thelocactus	SEED	Sow in a temperature of 21–24°C.	February to May	Globular cacti rarely producing offsets. Best as greenhouse plants in full light.

PLANT NAME	METHOD OF PROPAGATION	TREATMENT REQUIRED	TIME OF YEAR PROPAGATED	CULTURAL NOTES
Thrixanthocereus	SEED	Sow in a temperature of 21–24°C.	February to May	Hairy columnar cacti closely related to *Cephalocereus* and requiring the same conditions.
Titanopsis	SEED	Sow in a temperature of 21–24°C.	February to May	Succulent plants looking like rock fragments. Keep dry January to June cool but frost-free in winter.
Weingartia	SEED	Sow in a temperature of 21–24°C.	February to May	Globular cacti with swollen roots. A cold greenhouse should suffice for these fairly hardy plants.
Werckleocereus	SEED	Sow in a temperature of 21–24°C.	February to May	Trailing or climbing epiphytic cacti for a warm moist greenhouse in slight shade. Spray cuttings over occasionally until rooted.
	STEM CUTTINGS	Make cuttings from sections of young shoots and place on peat/sand compost to root.	May to September	
Wilcoxia	SEED	Sow in a temperature of 21–24°C.	February to May	Tuberous-rooted cacti with weak cylindrical stems.

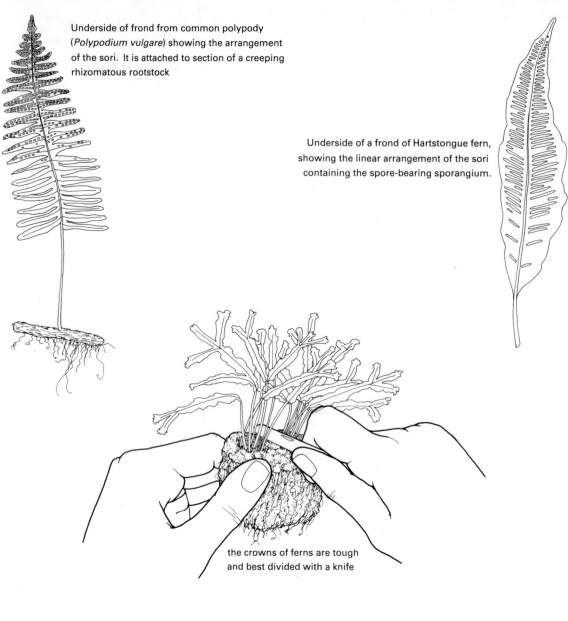

Underside of frond from common polypody (*Polypodium vulgare*) showing the arrangement of the sori. It is attached to section of a creeping rhizomatous rootstock

Underside of a frond of Hartstongue fern, showing the linear arrangement of the sori containing the spore-bearing sporangium.

the crowns of ferns are tough and best divided with a knife

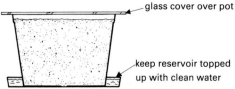

glass cover over pot

keep reservoir topped up with clean water

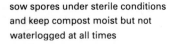

sow spores under sterile conditions and keep compost moist but not waterlogged at all times

A young fern growing from the prothallus

12 Ferns

Ferns find a place as foliage plants in many situations in the garden, greenhouse or home and are especially useful as many grow best in those damp shady corners so disliked by most other plants.

Numerous composts are recommended for various ferns but as far as results are concerned there seems little to choose between them, so as a general guide I would pot terrestial types such as tree ferns in a compost of 3 parts (by volume) roughly chopped moss peat, 1 part loam and 1 part coarse sand with perhaps a little ground charcoal to keep the compost sweet. The epiphytes such as *Davallia* do need a little extra fibre, however, so to the above I would add 2 parts chopped sphagnum moss and 1 part leaf-mould which should suit for most. Very many ferns grow best when slightly potbound so do not be tempted to overpot.

There is also a wide variety of ferns suitable to greenhouse cultivation. An unheated north-facing conservatory is ideal for the display of hardy types but those of a more tender nature require heat to a greater or lesser degree.

Many tropical species require a very warm greenhouse with minimum temperatures of 21 °C in summer and only slightly less in winter. The heat is best provided by pipes rather than air heat as these are less drying. The atmosphere should be moist at all times and ventilation given only when temperatures are extreme. Plants from more temperate regions will be happy in a cool house heated to around 10–13 °C in winter. In all cases shade will be required from the hot sun in summer.

Propagation of ferns is usually by spores or some form of division. Full details of spore raising will be found in Chapter 2.

PLANT NAME	METHOD OF PROPAGATION	TREATMENT REQUIRED	TIME OF YEAR PROPAGATED	CULTURAL NOTES
Acrostichum	SPORES	Sow in a temperature of 18–21°C. Keep compost moist at all times.	March to June	Large tropical ferns for a warm greenhouse where they are best grown with pots stood in water.
Actiniopteris	SPORES	Sow in a temperature of 18–21°C. Sow on a compost of brickdust with a little loam mixed in.	March to June or as soon as ripe	Small ferns with palm-like fronds. Requires a warm greenhouse with a winter minimum of 16°C.
Adiantum (Maidenhair Fern; A. caudatum roots at frond tips to form new plants; makes good houseplant)	SPORES DIVISION (of rhizome-bearing species)	Sow in a temperature of 18–21°C. Divide rhizomes and replant according to species. Plant hardy types 23–30 cm apart.	March to June In spring	A. raddianum (syn. A. cuneatum) makes an ideal houseplant for a moist position free of draughts. A. pedatum is deciduous and hardy. Many other species for all positions.
Aglaomopha (A. meyeniana = Bear's Paw Fern)	SPORES	Sow in a temperature of 18–21°C.	March to June	Large ferns for a warm greenhouse. Grow on a stump or in a basket pegging rhizomes down on rough moss peat.
Alsophila	SPORES	Sow at 18–21°C. Prick out singly to encourage strong straight trunk (caudex).	March to June	Greenhouse tree ferns only requiring frost-free conditions in winter. Grow in shade and water freely in summer.
Aneimia (Flowering Ferns)	SPORES DIVISION (of plants forming crowns)	Sow in a temperature of 18–21°C. Replant divisions in small pots.	As soon as ripe or in early April March/April	Small ferns for a warm greenhouse or closed frame indoors. Need more light than many ferns. Keep foliage dry.
Angiopteris	DIVISION (offsets: 'scales' or plantlets)	Root fleshy scales from the base of a frond on silver sand under a covering of sphagnum moss in a moist propagator at 21–27°C.	February/March	Huge ferns for a warm greenhouse having a caudex 1.8 m tall and fronds from 1.8 m to 4.5 m in length when fully grown.
Asplenium (A. nidus = Bird's Nest Fern; A. bulbiferum = Mother Spleenwort bears plantlets on the fronds)	SPORES DIVISION (hardy species bearing crowns)	Sow at 10–16°C for hardy types; 18–21°C for more tender kinds. Divide and replant in pockets in the rock garden. Appreciates a little mortar rubble in compost.	As soon as ripe or in spring In spring	Many species for the garden or greenhouse. A. nidus makes a good houseplant in a 13 cm pot. A. trichomanes is hardy and makes a good edging plant in sun or semi-shade.
Athyrium (A. filix-femina = Lady Fern)	DIVISION (of crown)	Lift and divide overcrowded plants replanting divisions with crown level with the soil surface.	March/April	The Lady Ferns are hardy deciduous ferns preferring a moist deep soil. Can be raised from spores but these do not come true.
Blechnum (B. spicant = Hard Fern. B. penna-marina = Sea Feather)	SPORES DIVISION (of rhizome-bearing types)	Sow at 10–16°C for hardy types, 18–21°C for more tender kinds. Divide and replant according to the requirements of the species.	March to June March/April	Many species both hardy and for the greenhouse. B. spicant is a hardy evergreen growing 30–38 cm in height. All make good cut foliage for flower arrangements.
Brainea	SPORES	Sow in a temperature of 18–21°C. Transplant singly to encourage strong trunk.	March to June	Tree ferns for a warm greenhouse. Transplant carefully as roots are brittle.
Camptosorus (C. rhizophyllus = Walking Fern)	DIVISION (of natural layers)	Fronds often root at the tip and can be separated when young plants have formed.	March/April	Hardy ferns similar to and needing the same conditions as our own Hart's Tongue Fern.

PLANT NAME	METHOD OF PROPAGATION	TREATMENT REQUIRED	TIME OF YEAR PROPAGATED	CULTURAL NOTES
Ceratopteris (Floating Stag's Horn Fern)	SPORES	Sow at 18–21°C keeping compost moist and sinking pots with young plants in water so that their crowns are covered.	In early spring	Floating Ferns for a pool in a warm greenhouse or a heated aquarium with a water temperature of 18–20°C. Pots should be stood in water up to their rims. Fronds of some species are edible.
	DIVISION (offsets: bulbils)	Bulbils are produced on fronds and can be pegged down on moist compost to root.	At any time	
Ceterach (C. officinarum = Rusty Back Fern/Scale Fern)	SPORES	Sow in a temperature of 18–21°C.	March to June	Small ferns mostly best grown in a cool greenhouse, although C. officinarum is hardy for the rock garden.
Cheilanthes	SPORES	Sow at 18–21°C. Often grow from self-sown spores.	March to June	Small evergreen ferns for the cool greenhouse. Do not water foliage and grow near the glass.
Cibotium	SPORES	Sow at 18–21°C. Prick out singly to encourage strong trunk.	March to June	Large easily-grown tree ferns for a warm greenhouse. As with all tree ferns it is best grown pot-bound.
Coniogramme (C. japonica = Bamboo Fern; genus sometimes included in Gymnogramma)	SPORES	Sow in a temperature of 18–21°C.	March to June	Medium-sized evergreen ferns for the cool greenhouse. Make sure pots are well drained and foliage is kept dry.
	DIVISION (of rhizome)	Repot in small pots.	February/March	
Cryptogramma (Mountain Parsley Fern/Rock Brake)	DIVISION (of crowns)	Replant divisions in small pots and grow on in cold frame until established.	Early spring before growth begins	Small deciduous ferns for the rock garden. Hardy. Can also be raised from spores.
Cyathea	SPORES	Sow in a temperature of 18–21°C. Prick out singly.	March to June	Evergreen tree ferns for a frost-free greenhouse or outdoors in mild areas. Keep trunks and roots moist and do not overpot.
Cyrtomium (C. falcatum = Holly Fern)	SPORES	Sow in a temperature of 18–21°C.	March to June	One of the ideal houseplants. Evergreen and grows 15–30 cm in height.
Cystopteris (C. fragilis = Brittle Bladder Fern)	SPORES	Sow at 10–16°C in propagator or cold frame.	March to June	Small hardy ferns growing only 15–20 cm tall. Grow in rock garden or 13 cm pan in an alpine house. Deciduous.
	DIVISION (of crowns)	Replant where they are grown. C. bulbifera can also be increased by bulbils on fronds.	March/April	
Davallia (D. canariensis = Hare's Foot Fern)	SPORES	Sow in a temperature of 18–21°C.	March to June	Epiphytic ferns for a warm greenhouse where most will grow in 15 cm pans. Many genera require similar cultivation to Davallias and these include Humata, Leucostegia, Leptolepia, Microlepia, Ochropteris, Odontosoria, Scyphularia and Stenoloma.
	DIVISION (of rhizome-bearing species)	Separate sections of rhizome and peg down on compost surface to root. Rhizomes must not be covered with compost.	March/April	
Dennstaedtia	DIVISION (of rhizomes)	Divide into sections when repotting and pot singly in small pots.	March/April	Smallish ferns mostly requiring a very moist and warm greenhouse in which to grow.
Dicksonia	SPORES	Sow in a temperature of 18–21°C.	March to June	D. antarctica is the hardiest of the tree ferns for mild areas outdoors or frost-free greenhouse. Others require some heat.
	DIVISION (of rhizome-bearing types)	Divide when repotting and repot singly in small pots.	March/April	

PLANT NAME	METHOD OF PROPAGATION	TREATMENT REQUIRED	TIME OF YEAR PROPAGATED	CULTURAL NOTES
Didymochlaena	SPORES	Sow in a temperature of 18–21°C.	March to June	Small evergreen ferns similar to the Common Polypody but requiring a warm greenhouse.
Diplazium (sometimes included in Asplenium to which it is closely related)	SPORES DIVISION (offsets: bulbils)	Sow in a temperature of 18–21°C. Detach bulbils and peg down on a compost of sand/peat in a warm greenhouse.	March to June At any time	Medium-sized evergreen ferns mostly requiring a very warm, moist greenhouse in which to grow. *D. proliferum* produces bulbils.
Doodia	SPORES	Sow in a temperature of 18–21°C.	March to June	Mostly low-growing ferns for the cool greenhouse. Division though possible is not recommended.
Dryopteris (D. filix-mas = Common Male Fern; D. linnaeana = Oak Fern; species of Lastrea are syn. of Dryopteris)	SPORES DIVISION (of species with creeping rhizomes)	Sow in a temperature of 18–21°C or slightly less for hardy species. Replant according to the requirements of the species.	March to June March/April	A very large genus of some 1,200 species. Hardy and tender types. Grow hardy types in sun or semi-shade in moist soil. Good for planting by water.
Elephoglossum (E. crinitum = Elephant's Ear)	Spores DIVISION (of crown)	Sow in a temperature of 18–21°C. Divide and pot singly when repotting.	March to June February/March	Evergreen ferns for a warm greenhouse where they grow best with their pots stood in water except for *E. crinitum* which should be watered sparingly.
Hemionitis	SPORES DIVISION (offsets: plantlets on fronds)	Sow in a temperature of 18–21°C. Peg down on pots of epiphytic compost to root.	March to June During summer	Small ferns for a warm greenhouse. Best grown in small pans of gritty compost. Also suitable for Wardian cases and bottle gardens.
Hemiptelia	SPORES	Sow at 18–20°C and prick out singly when large enough to handle.	March to June	Tree ferns mostly requiring a very warm greenhouse in which to grow.
Hymenophyllum (Filmy Ferns)	DIVISION (of rhizome)	Divide and peg the rhizomes down on the compost surface. Use the epiphytic compost. They are very slow-growing.	March/April	Very small ferns requiring humid cool conditions many being ideal for bottle gardens growing best where condensation provides all the moisture they need.
Hypolepis	SPORES (easy to raise) DIVISION (of rhizome)	Sow in a temperature of 18–20°C. Divide and peg rhizomes to compost surface.	March to June February/March	Medium-sized ferns for warm greenhouses. Need some grit or small stones mixing in compost to give perfect drainage.
Leptopteris (includes Todea)	SPORES	Sow at 18–21°C. Young plants are slow to develop.	March to June	Evergreen ferns similar to filmy ferns requiring a cool damp slightly heated greenhouse though *L. intermedia* is hardy.
Llavea	DIVISION (of crown)	Divide and replant in small pots or amongst rocks in a greenhouse border.	March/April	Evergreen fern for a cool house in a well-drained position. Do not wet foliage.
Lygodium	SPORES DIVISION (of crowns of most species)	Sow at 18–21°C. Divide and replant according to needs of the species.	As soon as ripe or in spring Spring through summer	Climbing ferns best planted in the borders in a warm greenhouse in moist compost. Grow up posts, trellis etc.
Marattia (M. galicina = Horseshoe Fern)	LAYERING	Bulbils will form on basal scales of mature fronds if these are pegged down to root on moist sphagnum moss.	In spring	Large evergreen ferns for a warm greenhouse where they need plenty of water. Can also be raised from spores.

PLANT NAME	METHOD OF PROPAGATION	TREATMENT REQUIRED	TIME OF YEAR PROPAGATED	CULTURAL NOTES
Matteuccia (*M. struthiopteris* = *Ostrich Feather Fern/Shuttlecock Fern*)	DIVISION (of rhizome)	Replant at least 60 cm apart. Prefers a moist position.	March/April	Hardy ferns growing 90 cm in height. Ideal for waterside plantings. Can also be raised from spores.
Nephrolepis (*Ladder Ferns*)	SPORES	Sow at 18–21°C.	March to June	Medium-sized ferns for the heated greenhouse where they grow well in wall or hanging baskets. Grow in the compost for epiphytes.
	DIVISION (of rhizome and crown)	Separate where young plants are rising from the rhizome. Replant according to species.	March/April	
Notholaena (*Gold and Silver Ferns*)	SPORES	Sow in a temperature of 18–21°C.	March to June	Most species of these small ferns need to be grown in heat though *N. marantae* needs only winter protection from rain. Need more light and air than many ferns.
	DIVISION (of species forming crowns or rhizomes)	Divide and replant in small pots growing on according to species.	March/April	
Oleandra	DIVISION (of rhizomes)	Replant on piles of fibrous peat by pegging rhizomes down, or train up mossed stake.	During summer	Creeping or climbing ferns for a warm greenhouse. Need plenty of water during summer.
Onoclea (*O. sensibilis* = *American Sensitive Fern*)	DIVISION (of rhizomes)	Lift and detach sections and replant just below soil surface. Has an indefinite spread.	March/April	Hardy ferns making good ground cover plants for moist places especially at the sides of streams.
Onychium	SPORES	Sow in a temperature of 18–21°C. Prick out as shallowly as possible.	March to June	*O. siliculosum* makes a good houseplant or warm greenhouse. Keep crown above soil surface when repotting.
Osmunda (*O. regalis* = *Royal Fern*; *O. cinnamomea* = *Cinnamon Fern*)	SPORES	Sow at 10–16°C. Spores lose viability if kept for more than three days after collection unless deep frozen.	As soon as ripe	Hardy waterside ferns growing up to 1.8 m tall with a spread of 1.2–1.5 m. Some species require a warm greenhouse. These are the plants that supply the *Osmunda* fibre used in orchid cultivation.
	DIVISION (of crowns)	Lift and divide well separated crowns. Replant in sun or slight shade.	March/April	
Pellaea (*P. rotundifolia* = *Button Fern*; *P. atropurpurea* = *Purple Cliff Brake*)	SPORES	Sow in a temperature of 18–21°C. Spores are easily raised and though those bearing rhizomes can be divided, better plants can be raised from the spores.	March to June	Small ferns, many such as *P. rotundifolia* making good houseplants or for bottle gardens. Most require a little heat but *P. atropurpurea* and one or two others are hardy enough for the rock garden.
Phyllitis (*P. scolopendrium* = *Hart's Tongue Fern*; *P. hemionitis* = *Mule's Fern*)	SPORES	Sow in a temperature of 13–18°C.	March to June	Mostly hardy ferns for the rockery though *P. brasiliensis* needs a very warm greenhouse. Hart's Tongue Fern makes a good houseplant.
	DIVISION (of crowns or by bulbils)	Divide the crown or grow on bulbils which sometimes form at the base of the frond.	March/April	
Pityrogramma (*Trismeria species require similar cultivation*)	SPORES	Sow in a temperature of 18–21°C.	As soon as ripe	Most species are best in hanging baskets in a warm but dryish greenhouse. Keep water off foliage.
Platycerium (*Stag's Horn Fern*)	SPORES	Sow in a temperature of 18–21°C.	March to June	Epiphytic ferns making good houseplants grown on a wad of sphagnum moss attached to bark or cork.
	DIVISION (of root buds where these appear)	Divide and pot on when first fronds appear from root buds.	As they become ready	

PLANT NAME	METHOD OF PROPAGATION	TREATMENT REQUIRED	TIME OF YEAR PROPAGATED	CULTURAL NOTES
Polypodium (P. vulgare = Common Polypody, a hardy species good for the rock garden)	SPORES DIVISION (of rhizome)	Sow at 18–21°C for tender species slightly less for hardy. Divide and peg rhizomes down on compost surface to root.	March to June April/May	A large genus of small evergreen ferns with finely divided fronds. There are species for every form of fern cultivation from hardy to very tender.
Polystichum (P.. acrostichoides = Christmas Fern; P. aculeatum = Hard Shield Fern)	SPORES DIVISION (of crown)	Sow at 10–16°C or more for tender species. Replant according to species' requirements. 15–18 cm pot indoors, 60 cm apart outside.	As soon as ripe or in spring March/April	Medium-sized evergreen ferns very many of which are hardy for moist shady situations. The remainder need heated glass in varying degrees.
Pteridium (P. aquilinum = Common Bracken Fern)	SPORES	Sow in a temperature of 10–16°C.	March to June	Hardy deciduous medium-sized fern which tends to be invasive if too well suited. Not suitable for pots.
Pteris (Doryopteris is cultivated in a similar manner to Pteris)	SPORES DIVISION (of rhizomes or crown)	Sow at 18–21°C. Prick out in small groups. Divide and replant in 8 cm pots when repotting.	March to June March/April	Small evergreen ferns many ideal for the house as they do not mind a dry atmosphere and most only require frost-free temperatures.
Taenitis (T. blechnoides = Ribbon Fern)	DIVISION (of rhizome)	Divide when repotting and repot in small pots.	March/April	Medium-sized ferns for a warm moist greenhouse and deep shade.
Tectaria	SPORES	Sow in a temperature of 18–20°C.	As soon as ripe	Fairly large but easily-cultivated ferns for a warm moist greenhouse.
Trichomanes	DIVISION (of rhizomes)	Replant sections by pegging rhizomes down on to a mixture of peat and sand at the base of the object over which they will grow.	Spring through summer.	Delicate scrambling ferns many of which cling to and grow over rock or dead tree trunks. Need warm greenhouse conditions. Can also be raised from spores.
Vittaria	DIVISION (of crowns)	Divide and repot in a sand/peat compost.	Spring through summer	Medium-sized ferns with grass-like fronds best suited to warm moist greenhouses.
Woodsia	SPORES DIVISION (of crowns)	Sow in a temperature of 18–21°C. Divide and replant in rock crevices or small pots of gritty compost.	March to June During winter when dormant	Small tufted plants many hardy and ideal for the rock garden or alpine house.
Woodwardia (W. arelata = North American Chain Fern)	DIVISION (offsets: bulbils; or of crowns)	Replant in rich wet soil allowing plenty of space for growth. May take some time to become established.	March/April	Hardy ferns for waterside plantings where they grow 45–60 cm in height.

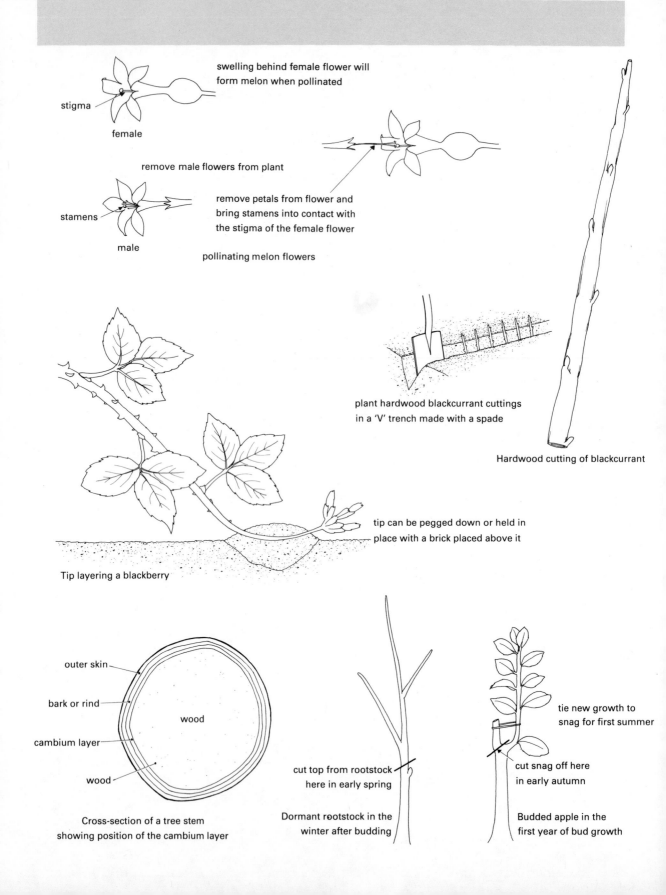

stigma

female

swelling behind female flower will form melon when pollinated

remove male flowers from plant

remove petals from flower and bring stamens into contact with the stigma of the female flower

stamens

male

pollinating melon flowers

plant hardwood blackcurrant cuttings in a 'V' trench made with a spade

Hardwood cutting of blackcurrant

tip can be pegged down or held in place with a brick placed above it

Tip layering a blackberry

outer skin

bark or rind

cambium layer

wood

wood

Cross-section of a tree stem showing position of the cambium layer

cut top from rootstock here in early spring

Dormant rootstock in the winter after budding

tie new growth to snag for first summer

cut snag off here in early autumn

Budded apple in the first year of bud growth

13 Fruit

The propagation of fruit-bearing plants ranges from the simple manipulation of natural layers, as in the strawberry, to the difficult 'double-working' when grafting incompatible pear stocks, but in all cases there is a fundamental requirement for cleanliness, both in stock and method of operation. Stock plants should be carefully selected for their health and vigour as nothing but disappointment will result from propagating unhealthy material.

By and large tree fruits grow best in southern areas and further north more care should be taken in selecting varieties to suit the climate, and in some cases it is necessary to ensure fruit production by growing under glass. Outside it is best to plant in sunny warm positions protected from cold winds in deep soil of good heart without being too rich. Soft fruit is more accommodating climatically and again should be planted in good deep soil. Many tree fruits are budded onto appropriate rootstocks, this being preferable to grafting in most cases, as it results in a better shaped tree. Buds are inserted from 8 to 30 cm above the soil surface but otherwise follow the same procedure as for roses (see Chapter 6). Rootstocks should be planted out 23 cm apart during winter to be budded the following July/August. If the bud fails to take make another attempt immediately, or graft the stock in the following March or April.

PLANT NAME	METHOD OF PROPAGATION	TREATMENT REQUIRED	TIME OF YEAR PROPAGATED	CULTURAL NOTES
Apples	BUDDING	Bud onto appropriate rootstock. Head stock back to 8–10 cm from the bud in spring using snag to tie new growth to create straight stem.	July/August	Apples will self-pollinate to a certain extent but are best grown with pollinators. Can be grown as 'family' trees. Rootstocks used to give particular results are as follows: Dwarfing M27, M9, M26. Semi-dwarfing MM106. Fairly vigorous M2, MM111. Very vigorous for forming large trees M25. New varieties are raised from 'sports' and seedlings. Seed should be sown as soon as ripe without drying out in a cold frame. Protect from frost for the first year.
	WHIP and TONGUE GRAFT	Cut stock back to 8–10 cm from the ground before making the graft. Plant out after 1 year. Distances apart: bushes 3.5–4.5 m; cordons 60 cm; espaliers 5.5 m; standards and half standards 6–7.5 m	February/March	
Apricots	BUDDING	Bud 15–23 cm above soil level onto apricot seedlings or onto the plum rootstocks Brompton, Common Mussel or St. Julien A.	July/August	Grow outside against a south or south-west wall or under glass in very cold areas. Apricots are self-pollinating outdoors but should be hand-pollinated under glass. Grow dwarf trees 3 m apart, larger 6–7.5 m.
	SEED (for rootstocks or raising new varieties)	Place in layers in peat/sand and place outside to stratify. Sow in rows in spring 5 cm deep, 23 cm apart.	As soon as ripe in auturnn	
Blackberry (also for Loganberries and Hybrid Berries)	TIP LAYERING	Bury tips of shoots 8–10 cm deep in soil where they touch or in a 13 cm pot dug in below shoot.	July/August	Separate new plants the following October or February. Replant from 2.5–3.5 cm apart against trellis or wire supports similar to those used for raspberries.
Blackcurrant	HARDWOOD CUTTINGS	Make cuttings 20–30 cm long, leaving lower buds on the stem to promote sucker growth. Plant in sheltered spot outdoors leaving 2 buds above the soil surface.	October/November	Grow in good moist soil in sun or slight shade planting 1.8 m apart. Blackcurrants are self-pollinating.
Cape Gooseberries (also known as Wonderberry/Golden Berry)	SEED	Sow 5 mm deep in a propagator at 21°C. Harden off and plant out when danger of frost has passed.	March	Grow outdoors or under glass in 20 cm pots. Grow outdoors in poor soil. Eat fruit raw or stewed, preserved or in pies.
Cherries	BUDDING	Bud onto cherry stocks Gean or Mazzard (Wild Cherry). Bud 15–23 cm from the ground or 1.8 m to create a standard tree.	July/August	Often large trees 9 × 9 m but dwarf trees are now possible on the newly introduced 'Colt' rootstock. With the exception of 'Stella', sweet cherries need pollinators, and as many varieties are incompatible care is needed in making the selection.
Cobnuts and Filberts	LAYERING or DIVISION (suckers)	Peg down 2-year-old branches and separate after 1 year. Plant out in nursery beds 25 cm apart for 3 years. Suckers can be treated in a similar manner.	October/November	Choose a site sheltered from east and north-east winds in well-drained positions. Grow as bushes planting 1.8 m–2.5 m apart. Self-pollinating.
Figs	SEMI-HARDWOOD HEEL CUTTINGS	Root 10–15 cm cuttings 1 to an 8 cm pot in a cold frame.	August/September	Best grown as bushes or fan-shaped trees against a south wall in mild areas, elsewhere under glass. Grow 4.5–5.5 m apart in poor soil and prune established roots to encourage fruiting. Self-pollinating.
	HARDWOOD CUTTINGS	Make 20–30 cm cuttings from ripe 1-year-old shoots. Remove buds from lower two-thirds and plant 15 cm deep, 15 cm apart in warm spot outdoors.	October/November	

PLANT NAME	METHOD OF PROPAGATION	TREATMENT REQUIRED	TIME OF YEAR PROPAGATED	CULTURAL NOTES
Gooseberry	HARDWOOD CUTTINGS	Make 20–30 cm cuttings removing buds from lower two-thirds before inserting 15 cm deep, 15 cm apart, in warm spot outdoors. Prune after 1 year; plant out after 2.	November	Grow in well-drained positions in sun or semi-shade planting 1.8 m apart. Can also be grown as cordons planting 40 cm apart.
Grapes	EYE CUTTINGS	Root in a propagator at 24°C. When well rooted pot on in 15 cm pots in a cooler and lighter position.	December/January	Grow outdoors against a south facing wall in mild areas planting 1.2 m apart. In other areas grow under glass. Because of its susceptibility to some insect pests *V. vinifera* is often grafted to resistant stocks.
Medlars	BUDDING	Bud named varieties onto seedling Medlars, Quince or Hawthorn.	July	As much grown as an ornamental tree as a fruit tree. The fruits can be eaten raw when ripe or used in jellies and preserves. Grows 6–7.5 m in height and as much across.
	SEED (suitable for ornamentals, rootstocks and raising new varieties)	Sow in a cold frame and pot on in 8 cm pots or in nursery beds outdoors when large enough to handle.	September/October	
Melons	SEED	Sow singly in small pots in a propagator at 24°C.	January/February	Grow in warm, moist greenhouse planting 45 cm apart on mounds of compost. Requires artificial pollination.
Mulberries	HARDWOOD CUTTINGS	Root 30 cm cuttings in a sand/peat compost in a cold frame or in a warm shady spot outdoors planting so only 2 buds remain above the surface.	September/October	Make good ornamental trees 6–7.5 m in height. Use fruits in tarts, preserves or wine-making. Grow against a south wall in very cold areas.
Peaches and Nectarines	SEED (Fruiting plants can be obtained from seed)	Place out to stratify in layers between peat/sand then sow 1 to a 13 cm pot in a cold frame in March.	October/November	Grow as wall trained plants 4.2 m apart against a south wall or under glass in all but the mildest areas. Self-pollinating outdoors need artificial pollination under glass. 1-year old seedling peach can also be used as a rootstock.
	BUDDING (of named varieties)	Bud 15–23 cm above soil surface on plum rootstocks St. Julien A for small trees or Common Mussel for medium trees.	July/August	
Pears *(Some varieties are incompatible with Quince and need to be double-worked; use 'Beurre Hardy' as the intermediate scion)*	BUDDING	Bud onto rootstock Quince A for vigorous trees, Quince C for smaller trees in good soil.	July/August	Grow in warm positions out of frost pockets. Plant out as for apples. With the exception of 'Conference' which is partially self-fertile pears need cross-pollinators and these need to be chosen with care to ensure compatibility. Some varieties best double-worked are: 'Clapp's Favourite', 'Bristol Cross' and 'William's Bon Chretien' (Bartlett).
	WHIP and TONGUE GRAFTING	Cut stock down to 8–10 cm from the soil surface. Make the graft using retarded scions. Plant out after 1 year.	March/April	
Plums, Gages and Damsons	BUDDING	Bud onto Myrobalan B. For vigorous trees (all except gages), Brompton or Common Mussel for medium trees and St. Julien A for semi-dwarfs.	July/August	Like apples and pears plums need cross-pollinators with a few exceptions such as damsons which are self-fertile. Plant standards 6–7.5 m apart, bush and half-standards 4.5–5.5 m apart and fan-shaped 4.5 m apart.
	DIVISION (suckers)	Separate suckers and plant as appropriate (damsons only).	During winter when dormant	

PLANT NAME	METHOD OF PROPAGATION	TREATMENT REQUIRED	TIME OF YEAR PROPAGATED	CULTURAL NOTES
Quince *(Planting distances: standards 6 m, bushes and half-standards 3–3.5 m)*	LAYERING	Peg lower branches down to soil and separate when rooted.	Late summer/early autumn	Low round-headed trees requiring moist soil and ideal for the waterside, but for fruit production grow against a south wall in cold areas. Self-pollinating. Use fruit in preserves and jellies.
	HARDWOOD CUTTINGS	Root 23–30 cm cuttings in a warm shady place outdoors.	October/November	
	STOOLING	This method is used when propagating for rootstocks.		
Raspberries	DIVISION (suckers)	Carefully separate and replant cutting canes down to within 15–20 cm of the ground.	In early autumn	Grow in sun or slight shade 45–60 cm apart in rows 1.5 m apart.
Rhubarb	DIVISION (of crowns)	Divide into single crowns and replant 90 cm apart.	March	Not strictly a fruit but the stems are used as a dessert in pies, stewed or in preserves, so it is included here. Grow in full sun in good soil and do not harvest in first year of growth. Plant out in autumn.
	SEED	Sow in slight heat and plant out when ready, or sow 2.5 cm deep in drills outdoors, thinning to 15 cm apart.	March/April April/May	
Red and White Currants	HARDWOOD CUTTINGS	Make cuttings 30–38 cm long and remove all buds except top 4–5 before inserting 15–20 cm deep in shady spot outdoors. Lift after 1 year and remove topmost roots and grow on for 1 more year in nursery beds.	October/November	Grow in sunny positions planting bushes 2 m apart, cordons 40–45 cm apart. Self-pollinating. Use fruit in pies and preserves.
Strawberries	LAYERING (natural)	Peg runners down in pots or in rows of the bed. Separate when well rooted.	Late spring/early summer	Grow in sun or slight shade planting 45 cm apart in rows 75 cm apart. Can also be grown in 'tower pots' or under glass.
Walnuts	SEED	Place out to stratify in layers between peat/sand. Sow 8 cm deep, 15 cm apart, in prepared bed outdoors in March.	As soon as ripe	Large slow-growing trees taking up to 20 years to mature. Seedlings will fruit but will be variable in results. Self-pollinating but advisable to have more than 1 tree. Plant 12–15 m apart in mild areas only as Walnuts are susceptible to spring frosts. The grafting of walnuts is difficult, needing much experience before it is successful.
	DOUBLE WEDGE GRAFTING (for named varieties)	As rootstocks, seedlings are lifted in November and potted in 10–13 cm pots being placed in slight heat in February. After grafting they are placed in a closed frame at 21–24°C until a union has been made.	March/April	

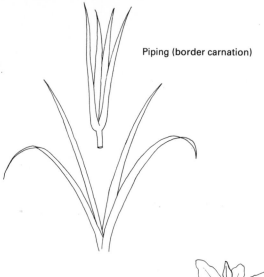

Piping (border carnation)

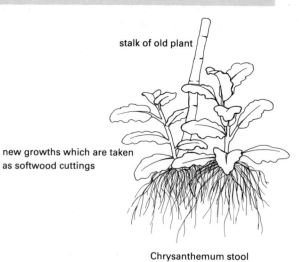

stalk of old plant

new growths which are taken
as softwood cuttings

Chrysanthemum stool

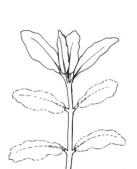

chrysanthemum cutting
prepared for planting by
removal of lower leaves

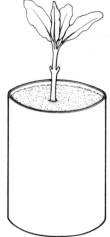

dip prepared cuttings in a
hormone rooting compound

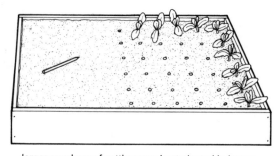

larger numbers of cuttings are best planted in boxes

Rosa canina

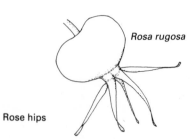

Rosa rugosa

Rose hips

taking a basal cutting
from a dahlia tuber

14 General Garden Plants

Contained in this chapter are those few types of plants which by their diversity and beauty have become generally accepted as garden 'specialities', both by the home gardener and plant breeder. For simplicity I have also included here a few plants of lesser importance but which are also grown aside from the main patterns of the garden for cut flower purposes etc (for example, stocks and sweet williams).

Detailed cultivation details would take much space and as with many other portions of this book I have to be content with merely starting the process off at the propagation stage. Perhaps with garden specials more than many other plants, attention to good cultivation will repay dividends and results will be generally determined by the amount of effort and expertise shown by the grower. This being so I would advise the enthusiastic beginner to consult one of the many excellent books written for his or her own chosen speciality.

PLANT NAME	METHOD OF PROPAGATION	TREATMENT REQUIRED	TIME OF YEAR PROPAGATED	CULTURAL NOTES
Carnation *(Dianthus caryophyllus)* *Perpetual-flowering* *types*	SOFTWOOD CUTTINGS	Root 10–13 cm cuttings taken from lower sideshoots in sand/peat compost in a propagator at 13–18°C or under mist with a bottom heat of 21°C. Set cuttings 1.2–2.5 cm apart.	October through to February	Perpetual-flowering carnations require greenhouse cultivation with a 7°C winter minimum. Pot cuttings into 8 cm pot when rooted and on to a 15 cm pot for final development.
Border Carnations and Pinks	SEED	Sow 5 mm apart in greenhouse or cold frame, or earlier in heat if desired.	April/May	Plant outdoors in late May 23–30 cm apart. Grow in sunny positions in good but not over-rich soil. Separate layers in September and plant out or pot to overwinter in cold frame.
	PIPINGS	Root in pots in a cold frame.	June to August	
	LAYERING (border carnations)	Use non-flowering shoots, cut a tongue and peg down in pots or border soil to root.	July	
Chrysanthemum *(modern large flowering* *cultivars are varieties of* *C. moriflolium, itself a* *complex hybrid)*	BASAL CUTTINGS	Take cuttings 5–8 cm in length. Late flowering large exhibition	December/January	Grow early-flowering types, sprays, poms, otley koreans and rubellums in full sun in the open garden in rich soil. Grow late-flowering types in large pots outdoors to house in a frost-free greenhouse after buds are formed but before autumn frosts begin.
		Ordinary late flowering types	January/February	
		Early flowering types, sprays, otley koreans etc. Root in a propagator at 10–16°C or under mist.	February to April	
Charm, Cascade and plants being raised for new varieties Pot Mums *(commercially produced* *by altering day length* *and treating with* *dwarfing chemicals)*	SEED	Sow in a temperature of 16–21°C. Seedlings will not come true.	February	After flowering cut plants down to 15–23 cm lift and plant in boxes to overwinter in cold frame or slightly heated greenhouse. Keep moist but not over-wet and start into growth in spring to provide cuttings. Harden off well before planting out in May.
	SOFTWOOD CUTTINGS	For natural dwarf effect root several cuttings to a pot, and grow on there to maturity. Most suitable varieties are those belonging to the Princess Anne family.	Late August/September	
Dahlia	DIVISION (of tubers)	Divide large tubers only cutting with at least 1 'finger' of tuber and a growth bud to each section.	In spring	Start tubers into growth in March by placing tubers in boxes of compost and watering. Grow on in slight heat. Plant out early in June when all danger of frost has passed. Plant out between 0.8 and 1.2 m apart depending on type. Store tubers in dry airy, frost-free place in winter.
	BASAL CUTTINGS	Make 8 cm cuttings cut below a node or with a heel of tuber attached. Root in a warm moist propagator.	April/May	
Lawns	SEED	Broadcast sow at a rate of 35–70 gm² for rye mixtures, 70–105 g/m² for finer types.	March/April or August/September	Choose seed mixtures to suite requirements. Those containing rye grass give the hardest wear. Prepare the seedbed thoroughly before sowing and protect from birds with black cotton. Give the first mowing when the seedlings are 8 cm high.
	TURFING	Lay turfs closely in staggered rows on a bed prepared as for seeds. Fill any gaps or depressions with riddled soil.	October/November or January/February	
Pelargonium *(geraniums: zonal types* *are hybrids of P. zonale;* *regal types are hybrids of* *P. domesticum)*	SOFTWOOD STEM CUTTINGS	Make 8 cm nodal cuttings and root in a peat/sand compost in a slightly heated propagator.	August/September	Easily-grown plants, ideal for the slightly heated greenhouse or as houseplants when in flower. Zonal pelargoniums are the type commonly used in summer bedding schemes. Lift and grow on in frost-free place for the winter.
	SEED (will not breed true)	Sow in a propagator at 18–21°C and pot on in slight warmth under glass.	January	

PLANT NAME	METHOD OF PROPAGATION	TREATMENT REQUIRED	TIME OF YEAR PROPAGATED	CULTURAL NOTES
Roses				Many kinds for garden decoration and cut flowers. Grow in sunny sheltered positions in good, preferably heavy soils. As well as being budded the stronger growing floribundas and hybrid teas can be propagated from cuttings as can climbers and ramblers. Planting distances: Bush types 60–90 cm apart, shrubs 1.2 m apart and standards 90 cm apart. Plant out between November and February.
	BUDDING	Bud onto the appropriate rootstock as outlined in Chapter 6.	Late June to August	
Rootstocks and Rosa species	SEED	Stratify hips in layers between sand/peat mixture outdoors and move to a cold frame in March and sow as normal.	As soon as ripe or in December/January	
Floribundas and Hybrid Teas	HALF-RIPE CUTTINGS	Make 23–30 cm cuttings of present year's growth and root in shady spot outdoors or in cold frame.	September	
Climbers and Ramblers	HARDWOOD CUTTINGS	Make 23–30 cm cuttings and root in a shady spot outdoors.	September	
Miniature Roses	HALF-RIPE CUTTINGS	Root small cuttings in a cold frame or greenhouse.	September	
Stocks *(derived from Matthiola incana)*	SEED (10 week stocks etc, HHA)	Sow in a temperature of 13–16°C. Grow on in cool conditions for easier selection of seedlings of 'selectable' types.	February/March	Scented flowers ideal for cut flowers indoors or under cool or cold glass and in some cases as summer bedding plants. Grow in rich well-drained soil in sun planting out 30 cm apart. Selectable types are grown to select preferred double flowers. Prick out the *light* coloured seedlings only.
	SEED (Brompton and East Lothian types, HB)	Sow in a cold frame, prick out and overwinter there, planting out in March or growing on to flower in 13–15 cm pots under glass.	June/early July	
Sweet Peas *(Lathyrus odoratus)*	SEED (soak 24 hours in tepid water before sowing)	Sow singly in deep fibre pots in a propagator at 16–18°C, or sow in the open ground 2.5 cm deep, 8–10 cm apart, and thin to 23 cm later.	September/October or February to March	Grow in good rich soil in sun or very slight shade. Overwinter pot-grown plants in a cold frame protecting in severe weather. Plant out 30 cm apart in March or April. Stake climbing varieties.
Sweet William *(Dianthus barbatus. A perennial best treated as a hardy biennial)*	SEED	Sow in boxes in a cold frame and prick out 13–15 cm apart in a prepared bed outdoors. Plant in final positions 20–30 cm apart in autumn.	Late April/May	Many strains and varieties ideal for cut flowers, as border plants or for bedding. Some strains are annual being sown where they will flower in spring. Can also be grown from half-ripe cuttings in July if desired.

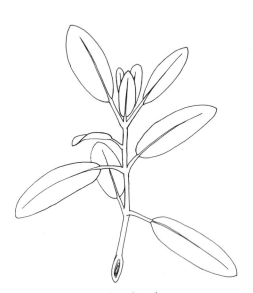

Heeled cuttings (sage)

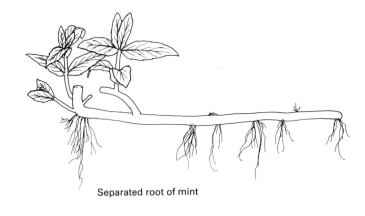

Separated root of mint

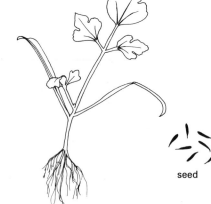

seed

Chervil seedling

parsley seedlings after germination

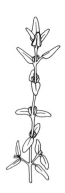

softwood cutting (thyme)

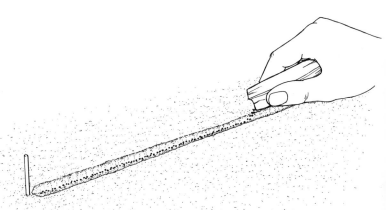

sowing herb seed in drills

15 Herbs

Many herbs can be grown in odd spots in the garden where there is good soil and where they will receive plenty of sun, but for best results they are better placed in a small plot of their own, preferably near the kitchen where they are handy for the cook. There is a wide variety to choose from and a well-planned herb garden can be beautiful as well as practical.

When propagating herbs vegetatively try and choose the parent with the best characteristics to suit your particular purpose, as there is often a wide variation amongst seedlings. The uses to which herbs can be put are many, and as well as details of propagation I have tried to indicate briefly what these may be in the cultural notes.

When gathering leafy herbs for drying choose a fine dry day when the plants are just beginning to flower. Hang the leaves up in bunches in a warm shady place to dry out quickly or place in layers in small trays and dry in an airing cupboard or very cool oven. When dry store in airtight containers. Seeds are saved when ripe in much the same manner as when saving seed for propagation purposes as described in Chapter 2.

PLANT NAME	METHOD OF PROPAGATION	TREATMENT REQUIRED	TIME OF YEAR PROPAGATED	CULTURAL NOTES
Agrimony *(Agrimonia eupatoria, HP)*	SEED (germination erratic) SOFTWOOD CUTTINGS	Sow shallowly in drills 30 cm apart outdoors. Root short cuttings in a cold frame.	March to May Spring or autumn	Medicinal. Flowering tops used in Agrimony tea. Sow thickly and thin to 25 cm apart. Grows 30–60 cm tall.
Angelica *(Angelica archangelica, HP)*	SEED (fresh seed gives best results)	Sow in open ground in slight shade and thin or transplant to 60 cm apart.	Spring or August/September	Culinary. Use leaves in salads, thin stems for candying. Makes large plant up to 1.5–1.8 m in height.
Anise *(Pimpinella anisum, HA)*	SEED	Sow in the open ground and thin to 30 cm apart. Can also be sown earlier under glass to be planted out in May.	April/May	Culinary and medicinal. Use leaves for flavouring soups, stews and salads. Seeds also used for flavouring many dishes and as a tisane aiding digestion. Grows 45 cm in height.
Basil (Sweet) *(Ocimum basilicum, HHA)*	SEED	Sow in pots in a cold frame and plant out in sunny spot in late May 30 cm apart.	March/April	Culinary. Use fresh or dried leaves in salads, soups and sauces. Use sparingly. Grows to 60 cm in height.
Bay *(Laurus nobilis; hardy evergreen tree)*	HALF-RIPE CUTTINGS	Root in a cold frame or propagator and overwinter under glass. Plant out in October of following year in tubs or open ground in sunny spot.	August/September	Culinary. Use dried leaves to flavour soups and stews and in bouquet garni. Grows into large tree but can be clipped to form a bush of any size.
Bergamot *(Monarda didyma, HP)*	DIVISION	Divide and replant 45 cm apart in damp position in slight shade. Grows 60–90 cm in height.	March	Culinary and medicinal. Leaves and flowers used fresh to flavour drinks and salads or dried in pot pourri. Can also be raised from seed.
Borage *(Borago officinalis, HA)*	SEED (prolific at self-seeding)	Sow in open ground in rows 45 cm apart and thin to 38 cm apart.	March to May	Culinary. Use young cucumber flavoured leaves to flavour drinks and salads. Flowers can be candied. Grows 30–60 cm in height.
Calamus *(Acorus calamus = Sweet Flag, HP)*	DIVISION (of rhizomatous roots)	Replant in wet soil at margins of ponds and streams 30 cm apart.	March	Culinary, medicinal and as ornamental plant. Roots can be candied or dried for medicinal use. Leaves will impart a cinnamon flavour to custards and creams.
Caraway *(Carum carvi, HB)*	SEED (autumn-sown seed gives best results)	Sow shallowly in open ground and thin to 15 cm, then 30 cm apart as they grow.	March/April or in autumn	Culinary and medicinal. Seed used for flavouring cakes etc. Fresh leaves can be used in salads and roots can be boiled.
Catnip *(Nepeta cataria, HP)*	SEED HALF-RIPE CUTTINGS DIVISION	Sow in open ground and thin to 15–20 cm apart. Root in a cold frame. Replant in the open ground.	March to May or September/October June/July March	Medicinal, and at one time culinary. Grows 60 cm in height and makes a good bee plant. The scent of crushed leaves is irresistible to cats.
Chamomile *(Anthemis nobilis, HP)*	SEED HALF-RIPE CUTTINGS DIVISION	Sow in open ground or in slight heat in February or March. Root in propagator or cold frame. Replant in the open ground preferably in well-drained sandy soil.	April July/August March	Medicinal. Flowers used to make chamomile tea and plants are also distilled for their oil. Thin or plant out 30 cm apart for cropping and 10–15 cm to create a chamomile lawn.

PLANT NAME	METHOD OF PROPAGATION	TREATMENT REQUIRED	TIME OF YEAR PROPAGATED	CULTURAL NOTES
Chervil *(Anthriscus cerefolium, HA)*	SEED (surface sow)	Sow outdoors in rows 30 cm apart and thin to 15 cm apart. Do not transplant.	March/April and August/September	Culinary. Use fresh leaves in salads, soups and sauces. Can also be grown in slight heat under glass for winter use.
Chicory *(Cichorium intybus, HP)*	SEED	Sow outdoors in shallow drills 30 cm apart and thin to 20–30 cm apart.	Mid-June to mid-July or April to June to produce roots for forcing.	Culinary. Use blanched leaves in salads and force roots in winter to produce 'chicons'. Dried roots can be used as a coffee substitute.
Chives *(Allium schoenoprasum, HP)*	SEED	Sow outdoors in shallow drills and thin to 15 cm apart.	March	Culinary. Use leaves chopped finely for flavouring salads and cream cheeses. Grows 15 cm in height and makes a good edging plant.
	DIVISION (offsets)	Replant in open ground 15 cm apart.	March	
Clary *(Salvia sclarea, HA or HB)*	SEED	Sow outdoors in shallow drills or in a cold frame in May/June. Thin or plant out 30 cm apart. Grows 60 cm tall.	March/April	Medicinal and culinary. Try fresh leaves fried in batter. Oil is distilled from the flowers to be used as a fixative in perfumery.
Comfrey *(Symphytum officinale, HP)*	DIVISION (root sections)	Plant outdoors 75 cm apart in slight shade.	March	Medicinal. Bruised leaves applied to sprains reduce swelling. Roots and leaves used as infusion for chest ailments.
	SEED	Sow outdoors in open ground. Grows to 1.2 m in height.	Spring	
Coriander *(Coriandrum sativum, HA)*	SEED	Sow in boxes under glass or outside in shallow drills in April and thin or plant out in May 38 cm apart.	February/March	Culinary. Use crushed seed as a seasoning in meat dishes. Grows 45–60 cm in height.
Corn Salad *(Lamb's Lettuce, Valerianella olitoria, HA)*	SEED	Sow outdoors in shallow drills 20 cm apart. Thin to 15 cm apart.	August/September and through spring months	Culinary. Thinnings are eaten in salads. Sow at intervals for succession.
Costmary *(Chrysanthemum [Tanacetum] balsamita, HP. Also known as Alecost)*	DIVISION	Replant in warm spot outdoors 60 cm apart. Will grow to 60–90 cm in height.	Spring or autumn	Culinary and medicinal. Use spring leaves in salads. Bruised leaves are soothing when applied to bee stings.
Cress (Land) *(Lepidium sativum, HA)*	SEED	Sow in open ground in moist soil in shallow drills and thin to 23 cm apart.	March/April	Culinary. Use in salads as for water cress. Grows up to 45 cm in height.
Cress (Water) *(Nasturtium officinalis, HP)*	SEED	Sow shallowly in pots stood in water in a cold frame. Plant out 13–15 cm apart.	March to May	Culinary. Use fresh young shoots in salads. Best transplanted into shallow running water but will grow in a moist situation.
Cumin *(Cuminum cyminum, HA)*	SEED	Sow in warmth and plant out 23 cm apart in May after hardening off.	February/March	Culinary. Seeds used to flavour pickles, chutneys etc. Grow in a warm sunny spot.
Dill *(Anethum graveolens, HA)*	SEED (will self-seed)	Sow in drills 25 cm apart in a sunny position and thin to 20 cm apart.	March	Culinary. Young leaves are used to flavour soups and stews, seeds used in dill vinegar for pickling. Grows 90 cm in height.
Elder *(Sambucus nigra. Hardy tree growing up to 9 m in height)*	HARDWOOD CUTTINGS	Root in sheltered spot outdoors or in cold frame. Plant out following autumn.	October/November	Culinary. Flowers used for wines or try one sprig stewed with rhubarb. Berries are used for wine and chutneys.
Elecampane *(Inula helenium, HP)*	SEED	Sow in sheltered spot outdoors and thin to 25 cm apart.	April/May	Medicinal. Whole plant rich in inulin. Roots used to produce fructose used in cough medicines. Grows 1.5–1.8 m in height.
	ROOT CUTTINGS	Make short cuttings each containing a bud and root in a cold frame.	September	

PLANT NAME	METHOD OF PROPAGATION	TREATMENT REQUIRED	TIME OF YEAR PROPAGATED	CULTURAL NOTES
Fennel *(Foeniculum vulgare, HP)*	SEED	Sow in drills outdoors and thin to 38 cm apart.	April	Culinary. Use leaves to flavour fish dishes, stems sliced in salads. Grows between 60 cm and 1.2 m in height.
	DIVISION	Replant in open ground.	March as soon as shoots appear	
Finnocchio *(Florence Fennel;* *Foeniculum dulce, HA)*	SEED	Sow outdoors in shallow drills and thin to 30 cm apart.	March/April or September	Culinary. Bulbous roots served raw or cooked as a vegetable in July. Use leaves as for ordinary fennel.
Garlic *(Allium sativum, HP)*	DIVISION (offsets)	Divide bulbs (cloves) and replant offsets outdoors in rows 30 cm apart, planting 15 cm apart in the rows.	February to April	Culinary. Crushed cloves used for flavouring with caution! Plant bulbs just below the soil surface.
Hop *(Humulus lupulus, HP)*	SEED	Sow outdoors in shallow drills and thin to 30–45 cm apart. Can also be divided in spring.	April/May	Culinary and medicinal. A climbing plant whose female flowers are used in beer-making and when dried can also be used as a filling to a sleep inducing pillow. Young shoots can be eaten as a vegetable.
Horseradish *(Cochlearia armoracia,* *syn. Armoracia rusticana,* *HP)*	ROOT CUTTINGS	Root short cuttings in a cold frame.	January/February	Culinary. Roots are used raw to make horseradish sauce a condiment best served with beef. Grows to around 60 cm in height.
	SEED	Sow outdoors and thin to 30 cm apart. Grow in a deeply cultivated soil.	March/April	
Hyssop *(Hyssopus officinalis, a* *small evergreen shrub)*	SEED	Sow outdoors in shallow drills and thin to 30 cm apart.	April	Culinary. Use fresh young shoots to season salads and soups. An oil is also distilled from the flowering tops.
	HALF-RIPE CUTTINGS	Root in a sand/peat compost in a cold frame.	August	
Lavender *(Lavandula vera, hardy* *shrub)*	HALF-RIPE CUTTINGS	Root 10 cm cuttings in a cold frame making sure they are firm.	August	Herbal. Dried flowers used to make pot pourri and are also distilled for oil. *L. spica* is also used and there are many varieties and forms.
	HARDWOOD HEEL CUTTINGS	Root 18–20 cm cuttings outdoors or in a cold frame.	October/November	
Lemon Balm *(Melissa officinalis, HP)*	SEED	Sow shallowly in drills outdoors and thin to 30 cm apart.	March/April	Culinary. Dried leaves are used in poultry stuffing and to impart a lemon flavour to Indian tea. Grows 60 cm in height.
	DIVISION	Replant in the open ground.	March	
Lemon Verbena *(Lippia citriodora,* *deciduous shrub or small* *tree)*	SOFTWOOD CUTTINGS	Root short cuttings in sand/peat compost in warm propagator. Grow on in frost-free conditions until established.	July	Herbal/medicinal. Leaves used to make Verbena tea and as a scent in soap. Rather tender so grow against a warm wall in colder areas.
Lovage *(Levisticum officinale,* *HP)*	SEED	Sow in a prepared bed outdoors and transplant to 6 cm apart.	March to May or in autumn when ripe	Culinary and medicinal. Use leaves as seasoning in salads and soups. Root is used medicinally by herbalists. Grows up to 1.5 m in height.
	DIVISION	Replant in the open ground.	Autumn or in early spring	
Marjoram (Pot) *(Origanum onites, HP)*	SEED	Sow in slight heat and plant out 23 cm apart in May.	February/March	Culinary. Use fresh or dried leaves gathered July to September in stuffings for poultry. Grows 30 cm in height.
	LAYERING	Layer shoots in the open ground as soon as they are long enough to peg down.	March onwards	

PLANT NAME	METHOD OF PROPAGATION	TREATMENT REQUIRED	TIME OF YEAR PROPAGATED	CULTURAL NOTES
Marjoram (Sweet) *(Origanum marjorana HHA or HHP)*	SEED	Sow outside in drills 30 cm apart or in slight heat inside hardening off before planting out in May 23 cm apart.	April/May February/March	Culinary. Use leaves and flowers fresh or dried to impart a sweet spicy flavour to food or as an ingredient in Bouquet garni and pot pourri. Grows 60 cm in height.
	BASAL CUTTINGS	Root short cuttings in a cold frame and protect from frost.	July/August	
Mint *(Mentha spicata = Spearmint; M. piperita = Peppermint; HP)*	DIVISION	Replant outdoors 15 cm apart.	March	Culinary and medicinal. Spearmint used to make mint sauces etc. The tops of Peppermint are used to produce Peppermint oil. Grows 30–60 cm tall.
	SOFTWOOD CUTTINGS	Root in the open ground or in a cold frame.	August	
	SEED	Sow in drills 23 cm apart outdoors.	March to May	
Mullien *(Verbascum thapsus, HB, also known as Aaron's Rod or Blanket Herb)*	SEED	Sow in a prepared bed outdoors or a cold frame and plant out in September 60 cm apart.	March to May	Medicinal. An infusion made from the leaves relieves pulmonary congestion. Strain through muslin to remove leaf hairs which are irritants. 1–1.2 m in height and makes a good border plant.
	ROOT CUTTINGS	Root in pots in a cold frame.	Autumn or winter	
Nasturtium *(Tropaeolum majus, HHA)*	SEED	Sow in open ground and thin as required. For use as herbs choose compact varieties.	April/May	Culinary. Use leaves and flowers in salads, unripe seed pickled as capers. Leaves can be dried and crumbled for seasoning.
Parsley *(Carum petroselinum, HB)*	SEED (soak in warm water for 12 hours before sowing)	Sow outdoors in rows 30 cm apart and thin to 15 cm apart. Grows 15–23 cm in height.	March/April and again in June/July	Culinary. Use fresh leaves as garnishing, flavouring and in sauces. Seed may take up to 8 weeks to germinate.
Pot Marigold *(Calendula officinalis, HA)*	SEED	Sow thinly in drills 45 cm apart and thin to 30 cm apart. Grows up to 90 cm in height.	March to May	Culinary and medicinal. Use young leaves and flowers in salads and dried flowers in soups and as colouring. Also used as a tisane.
Purslane *(Portulaca oleracea, HHA)*	SEED	Sow in slight warmth and plant out late May 23 cm apart. Can also be sown in drills outdoors in sunny position.	March/April May	Culinary. Use young fresh leaves and shoots in salads. Sow at intervals for succession. Grows 15 cm in height.
Rose *(Rosa damascena = Damask Rose)*	HALF-RIPE CUTTINGS	Root in a sheltered spot outdoors. Plant out in the following autumn 60 cm apart. Damask rose grows 1–1.5 m in height.	September	Medicinal, herbal aand culinary. Hips of *R. canina* and others are rich in vitamin C and can be made into jellies etc. Damask rose is used to produce an oil and petals can also be candied.
Rosemary *(Rosmarinus officinalis, a hardy evergreen shrub growing 1–1.2 m in height)*	SEED	Sow in shallow drills outdoors and transplant to nursery beds 15 cm apart.	April	Medicinal, herbal and culinary. Use leaves as a seasoning with chicken and fish. Dried leaves used in pot pourri and also in many medicines.
	HALF-RIPE HEEL CUTTINGS	Root in a shady cold frame and overwinter there.	August	
Rue *(Ruta graveolens, HP)*	SEED	Broadcast sow outdoors and thin to 38 cm apart.	March/April	Medicinal. Leaves are used in rue tea or as a medicine to cure croup in poultry. Grows 45–60 cm tall.
	SOFTWOOD CUTTINGS	Root in a shady cold frame and plant out in autumn.	March/April	
Saffron *(Crocus sativus)*	DIVISION (offsets; corms)	Replant in open ground 5–8 cm deep, 15 cm apart.	As soon as foliage dies down	Culinary. Use orange/red stigmas to produce a colouring agent and spice used in confectionery.

PLANT NAME	METHOD OF PROPAGATION	TREATMENT REQUIRED	TIME OF YEAR PROPAGATED	CULTURAL NOTES
Sage *(Salvia officinalis, HP)*	SEED	Sow outdoors in rows 45 cm apart and thin or transplant to 38 cm apart.	April	Culinary and medicinal. Leaves used to season stuffings, sausages etc. Infusion of leaves can be used as a gargle to relieve sore throats. Grows 30–60 cm in height.
	HALF-RIPE HEELED CUTTINGS	Root in pots in a cold frame and overwinter there.	May to September	
Salad Burnet *(Poterium sanguisorba, HP)*	SEED	Sow 13 mm deep in drills, 23 cm apart, and thin to 23 cm.	April	Culinary. Use young shoots in salads and as a seasoning in stews and soups. Grows to 38 cm in height.
	DIVISION	Replant 23 cm apart in open ground.	October/November	
Savory (Annual Summer) *(Satureia hortensis, HA)*	SEED	Sow in slight heat and plant out in May 20 cm apart. Can also be sown outdoors in shallow drills 30 cm apart.	March April	Culinary. Use fresh leaves sprinkled over salads, and in cooked dishes of pea, bean and mushroom. Grows 30 cm in height.
Savory (Perennial Winter) *(Satureia montana, HP sub-shrub)*	SEED	Sow outdoors in drills 30 cm apart.	April	Culinary. Uses similar to summer savory. Also makes an attractive shrub in the rock garden growing 30–38 cm in height.
	SOFTWOOD CUTTINGS	Root in a cold frame and plant out 45 cm apart.	May	
Sorrel (French) *(Rumex scutatus, HP)*	SEED	Sow thinly in shallow drills and thin to 30 cm apart.	Late March	Culinary. Young leaves used for flavouring salads, also in soups and sandwiches. Can also be cooked as a spring green vegetable. Grows 30–60 cm in height.
	DIVISION	Replant in the open ground and remove flowers as they form.	March/April	
Sweet Cicely *(Myrrhis odorata, HP)*	SEED	Sow in drills outdoors and thin to 45–60 cm.	March/April	Culinary. Fresh leaves are used in salads or boil the roots as a vegetable. Grows up to 60 cm in height. The root is also used medicinally as a tonic.
	ROOT CUTTINGS	Root in a cold frame.	February/March	
	DIVISION	Replant where they are to grow.	Spring or in autumn	
Tansy *(Tanacetum vulgare, HP)*	SEED	Sow outdoors and thin to 30 cm apart.	March/April	Culinary and herbal. Bitter leaves used to make tansy cakes. Flowers used to make a dye for wool. Grows 60–90 cm tall.
	DIVISION	Replant where they are to grow.	Autumn or early spring	
Tarragon *(Artemisia dracunculus, HP)*	DIVISION	Replant in an open well-drained spot 60 cm apart.	April/May	Culinary. Leaves used dried as a flavouring in vinegar, preserves and marinades. Can also be grown from seed when available.
	SOFTWOOD CUTTINGS	Root short cuttings in a cold frame. Grows 45–60 cm tall.	July/August	
Thyme *(Thymus vulgaris, HP)*	SEED	Sow in a cold frame and plant out 30 cm apart in September.	March/April	Culinary. Use shoots fresh or dried in bouquet garni or to flavour soups, meat and fish. Also a good plant for the rock garden.
	SOFTWOOD CUTTINGS	Root in a shady spot outdoors or in cold frame. 23–30 cm in height when established.	May	
Valerian *(Valeriana officinalis, HP)*	SEED (surface sow)	Sow outdoors fairly thickly and thin to 45 cm apart.	April	Medicinal. A nervine is produced from the roots, the smell of which is rather strong. Grows up to 1.2 m in height.
	DIVISION	Replant in moist well cultivated soil.	In autumn	
Violet *(Viola odorata, HP)*	SEED	Sow in pots and place outside to stratify and bring into a cold frame in March.	December/January	Culinary. Flowers can be crystallised as decoration in confectionery or used fresh in salads. There are many forms also suitable for garden decoration.
	DIVISION (rooted stolons)	Replant outdoors 30 cm apart.	April/May	

PLANT NAME	METHOD OF PROPAGATION	TREATMENT REQUIRED	TIME OF YEAR PROPAGATED	CULTURAL NOTES
White Horehound *(Marrubium vulgare, HP)*	SEED	Sow outdoors in rows and thin to 23 cm apart.	April/May	Medicinal and culinary. The juice from the boiled plant can be made into a candy to treat coughs or leaves can be used for flavouring but are rather bitter.
	DIVISION	Replant where they are to grow. 35–45 cm tall when fully grown.	Autumn or early spring	
Wintergreen *(Gaultheria procumbens, a low-growing evergreen shrub)*	SEED	Sow in a cold frame in a sand/peat compost and overwinter there.	In autumn	Medicinal. Leaves are distilled for the oil which is used in ointments. Makes a decorative plant in the rock garden.
	HALF-RIPE HEEL CUTTINGS	Root in a shady cold frame. Plant out in the following autumn.	July/August	

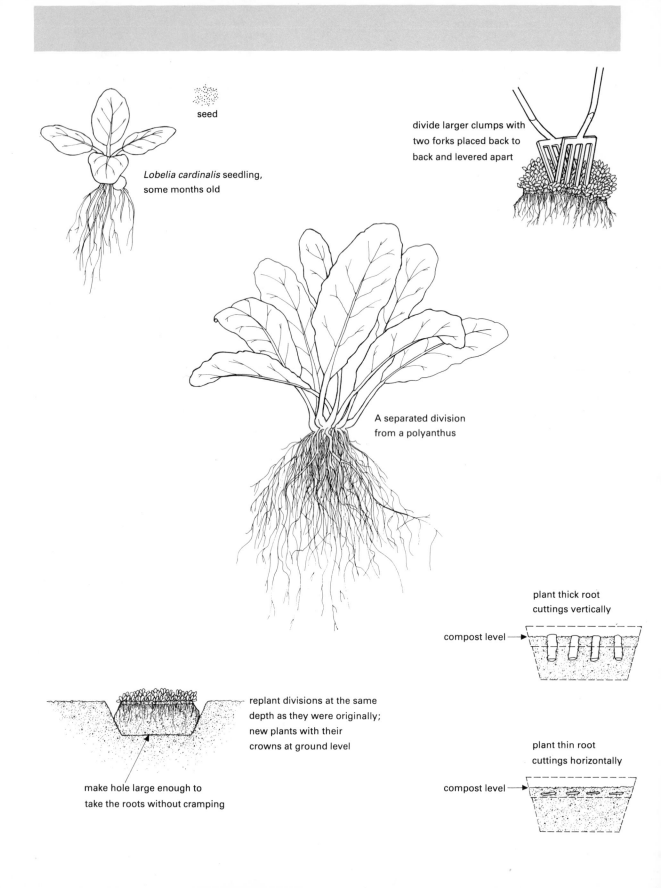

seed

Lobelia cardinalis seedling, some months old

divide larger clumps with two forks placed back to back and levered apart

A separated division from a polyanthus

plant thick root cuttings vertically

compost level

replant divisions at the same depth as they were originally; new plants with their crowns at ground level

make hole large enough to take the roots without cramping

plant thin root cuttings horizontally

compost level

16 Herbaceous and other Perennials

For ease of cultivation and maintenance there are few sections of the garden more amenable than the well planned and prepared mixed border containing a few shrubs and a selection of hardy perennial plants. To the purist, the real herbaceous border is one inhabited only by those perennials which lose all or most of their foliage in late autumn to remain dormant throughout the winter, but as I am far from a purist my borders are of the mixed type and it is on this basis that I have chosen the perennials for this chapter. As it happens the vast majority are herbaceous, however, and in the following text can be taken as such unless otherwise stated.

From a propagation viewpoint most perennials are easy, with division of the rootstock during their dormant period probably being the simplest and most speedy method, especially with those over-enthusiastic plants which need frequent lifting and dividing for their own good. Apart from specific varieties which should be raised vegetatively, most perennials can be raised from seed and usually flower in their second year or even in their first if sown early. In the text I have chosen to indicate sowing outdoors in late spring and early summer in most cases, but earlier sowings can be made under glass if you prefer, the seedlings being grown on as for HHAs. If, however, your covered space is at a premium in spring, then outdoor sowings are perfectly satisfactory. Seedlings are best thinned or transplanted in a nursery bed 15–23 cm apart for their first summer, to be planted in their permanent positions in autumn or early spring. For best effect plant in groups of one species or variety varying the number in a group from 3 to 5 depending on the size of border.

When creating a perennial border forward planning and preparation are essential. The size and shape is a matter for personal preference as is the type of plant effect one wishes to create, but the larger the border the more impressive the display. Borders can be made along the garden boundary or as 'islands' in a lawn or paved area, but aim for a width of at least 1.5–1.8 m, and choose plants to suit for height on the basis of a maximum plant height being equal to half the width of the border. Before planting the ground should be prepared to a good state of cultivation and the removal of every trace of perennial weed will save much future work.

PLANT NAME	METHOD OF PROPAGATION	TREATMENT REQUIRED	TIME OF YEAR PROPAGATED	CULTURAL NOTES
Acanthus (Bear's Breeches)	SEED	Sow in slight heat or in a cold frame.	March to May	Height 1–1.2 m. Flowering plants with interesting foliage. Grow in sun or slight shade and protect from frost in cold areas. Make good specimen plants.
	ROOT CUTTINGS DIVISION	Root in pots in a cold frame. Replant 30–60 cm apart where they are to grow.	Late autumn or winter In spring	
Achillea (Milfoil/Yarrow: herbaceous types)	DIVISION	Divide every 3 years and replant 30–45 cm apart as species require.	In spring	Many species for all places in the sunny border and others for the rock garden. A. filipendulina is the tallest up to 1.5 m and its flowers can be dried as 'everlastings'.
	SEED (fresh seed gives best results)	Sow in greenhouse or cold frame and overwinter late seedlings there growing them on in a nursery bed in spring.	February to May or when fresh ripe	
Aconitum (Monkshood/Wolf's Bane)	DIVISION (roots are poisonous)	Replant in sun or semi-shade 30–38 cm apart.	Autumn or spring	A number of species growing from 60 to 150 cm in height. Move carefully as they resent disturbance. Flower in late summer.
	SEED	Sow in shallow drills in a prepared bed and plant out in October.	April	
Actaea (Baneberry: A. spicata = Herb Christopher)	DIVISION (roots are poisonous)	Replant in moist shady positions about 30 cm apart.	In spring	Height 30–45 cm. Flowering/colourful fruited plants for the wild garden or among shrubs in moist soil. Berries and roots are poisonous.
	SEED (fresh seed gives best results)	Sow in drills in a prepared bed outdoors and plant out in autumn.	Autumn or spring when ripe seed is gathered	
Actinomeris (North American Sunflower)	SEED	Sow in a prepared bed outdoors and plant out in autumn.	April/May	Height 1.2–2.4 m. Tall flowering plants most suitable for a wild garden. Flowers in summer.
	DIVISION	Replant 50 cm apart.	In spring	
Adenophora (Ladybell/Gland Bellflower)	SEED	Sow singly in pots in a greenhouse or propagator and plant out when established or in spring. Resents disturbance once planted out.	In spring or when ripe seed is available	Perennials growing from 30–122 cm in height depending on the species. Flower from May to August. Do not divide. Grow in a sunny position.
Adonis (A. autumnalis = Pheasant's Eye HA: A. vernalis = Spring Adonis, HP)	DIVISION	Replant in open ground 15 cm apart.	In spring	Perennial flowering plant growing 30 cm high. Flowers in spring. A. amurensis flore pleno is the best.
	SEED	Sow in pots in a cold frame and plant out in spring in a nursery bed to grow on.	As soon as seed is ripe	
Agastache (A. mexicana = Mexican Bergamot)	SEED	Sow in slight heat and plant out when established.	February to June	Flowering aromatic perennial growing 60–75 cm high. Needs winter protection in all but mildest areas.
	DIVISION	Replant 45 cm apart.	In spring	
Ajuga (Bugle)	SEED	Sow outdoors or in boxes in cold frames.	In spring or autumn	Perennial carpeting plants with varieties of colourful foliage. Grow in sun for best effect but plants will also grow in shade.
	DIVISION	Replant 20–30 cm apart.	At any time	
	SOFTWOOD CUTTINGS	Root in pots in a cold frame.	August	
Alchemilla (A. mollis = Lady's Mantle)	SEED	Sow outdoors or in a cold frame and plant out in autumn.	Spring to early summer	A ground cover plant thriving in shady and semi-shaded spots. Is 45 cm in height when in flower.
	DIVISION	Replant 45 cm apart..	In spring	
Althaea (A. rosea = Hollyhock)	SEED	Sow in a cold frame: or in shallow drills in prepared bed outdoors.	April June/July	Height 1.8–2.4 m. Good plants for the back of borders. A number of types widens their use for summer bedding and as biennials. Plant out 38 cm apart.
	BASAL CUTTINGS	Root in slight heat and grow on under glass through the colder months.	At any time	
	ROOT CUTTINGS	Root and grow on in a cold frame.	During winter	

PLANT NAME	METHOD OF PROPAGATION	TREATMENT REQUIRED	TIME OF YEAR PROPAGATED	CULTURAL NOTES
Amsonia	DIVISION	Replant in slight shade 30–45 cm apart.	In spring	Height 30–90 cm depending on the species. Bear blue flowers during summer. Grow well beside shrubs.
	SOFTWOOD CUTTINGS	Root in pots in a cold frame.	July/August	
Anaphalis (Pearl Everlasting)	DIVISION	Replant 38–45 cm apart.	In spring	Height 30–60 cm. Densely-growing plants with grey foliage and white flowers. Grow in sun for best effects.
	SEED	Sow in a cold frame. Plant out in autumn.	In spring	
Anchusa (Bugloss/Alkanet)	ROOT CUTTINGS	Root in pots in a cold frame using thick sections of root.	Immediately after flowering	*A. azurea* flowers from 1–1.5 m in height, other species are less tall. Grow in sun planting out 30–38 cm apart. Flower from May to July.
	SEED	Sow in a temperature of 13–16°C and grow on in cold frame until established.	February/March	
Anemone (Windflower/Japanese Anemone)	ROOT CUTTINGS (A. lesseri and A. vitifolia)	Root in pots in a cold frame and grow on in pots until established.	During winter and early spring	Herbaceous perennial types grow to about 60 cm in height. Plant out in spring 30–38 cm apart. Grow in sun or slight shade. They flower in late summer and autumn.
	DIVISION (A. dichotoma and A. hupehensis)	Grow divisions on in pots in a cold frame.	In autumn	
	SEED (surface sow)	Sow in a propagator at 18–21°C.	February to May	
Anemonopsis	SEED	Sow and grow on in a cold frame.	In spring	Grows 60–90 cm in height. Bears its purple and lilac flowers in July.
	DIVISION	Grow on in pots in a cold frame until established.	In spring	
Anthemis (A. tinctoria = Golden Marguerite)	DIVISION	Replant in sunny spot 30–38 cm apart.	In spring	Flower from 38–60 cm in height. Flower freely and are good for cutting or border decoration.
	BASAL CUTTINGS	Root in cold frame or slight heat.	July/August	
	SEED	Sow in slight heat and harden off before planting out.	February/March	
Aquilegia (Columbine)	SEED	Sow in a propagator at 18–21°C or in a prepared bed outdoors in May.	March/April	Height 45–75 cm. Short-lived perennials for sun or slight shade. Flower in May and June. *A. alpina* is a rock garden plant.
	DIVISION (of named varieties)	Replant in the open ground 30 cm apart.	March	
Arnebia (A. echioides = Prophet Flower)	SEED	Sow in slight heat. Grow on in cold frame.	March or as soon as ripe	A plant for borders, rock gardens or dry walls growing 23–30 cm in height. Flowers in May. Grow in sunny well-drained positions.
	HEEL CUTTINGS	Root in sand/peat in slight heat and overwinter under glass.	In autumn	
	DIVISION	Replant 30 cm apart.	August/September	
Artemisia (A. absinthium = Wormwood)	DIVISION (herbaceous species)	Replant 38–45 cm apart in sunny positions.	In spring	Mostly grey-leaved foliage plants growing 60 cm to 1.2 m in height. Some species are used medicinally and as an ingredient of pot-pourri.
	HEEL CUTTINGS	Root in a sandy compost in a cold frame.	July/August	
Aruncus (A. sylvestris = Goat's Beard; syn. Spiraea aruncus)	SEED	Sow in a cold frame or in a prepared bed outdoors. Plant out in autumn 50 cm apart in sun or slight shade.	March to June	Height 1.2 m or more. Best seen as a specimen plant. Flowers in June and July. Prefers a moist soil.
Asclepias (A. incarnata = Swamp Milkweed; A. syriaca = Silkweed)	DIVISION	Replant 30–45 cm apart.	Autumn or early spring	Hardy perennial species grow from 60 cm to 1.2 m in height. They flower in July and make good border plants.
	SEED	Sow in slight heat or in a cold frame in April or May. Grow on in a nursery bed for the summer.	February/March	

PLANT NAME	METHOD OF PROPAGATION	TREATMENT REQUIRED	TIME OF YEAR PROPAGATED	CULTURAL NOTES
Aster (Michaelmas Daisies, especially A. novi-belgii)	DIVISION (most species)	Replant in the open ground from 30 to 38 cm apart in most cases.	Spring or autumn	Many species and varieties for all aspects of the flower border and for cutting. Flower in late summer. Divide strong growers fairly frequently.
	BASAL CUTTINGS	Root and grow on in pots in a cold frame.	In spring	
	SEED	Sow in a prepared bed outdoors.	May to July	
Astilbe (False Goat's Beard)	DIVISION	Replant in shady moist soil 30 to 38 cm apart.	April	Grow from 60 cm to 1.5 m in height depending on the variety. They flower from June to August. Keep roots moist at all times.
	SEED (slow germination)	Sow in pots in slight heat or in a cold frame.	In spring	
Astragalus (A. alopecuroides = Foxtail Milk Vetch)	SEED (slow to germinate)	Sow in a prepared bed outdoors, thin to 15 cm apart and plant out carefully in October. Resent disturbance once established.	In spring	Plants of medium height but variable. Grow upright types in the border, low-growing types in the rock garden. Plant in well-drained sunny spot.
Astrantia (Masterwort)	DIVISION	Replant 40 cm apart in moist soil.	March	Height 75–90 cm. Grow in sun or semi-shaded positions where they will flower in June, July and August.
	SEED	Sow in a moist shady prepared bed outdoors.	As soon as ripe or in spring	
Baptisia (False or Blue Indigo)	DIVISION (of woody rhizome)	Replant 60 cm apart in sunny well-drained positions.	Spring or autumn	B. australis grows to about 1.2 m in height and flowers in June.
	SEED	Sow in pots in a cold frame and plant out in autumn.	In spring	
Bellis (B. perennis = Common Daisy)	SEED	Sow in shallow drills in open ground. Thin and plant out in October 15–20 cm apart.	June	Some species suitable for the rock garden. B. perennis makes a good edging plant or for spring bedding when it is treated as a HB. Grows 15 cm tall.
	DIVISION (of named varieties)	Replant divisions where they will grow.	After flowering	
Bergenia (Bear's Ears/Pig Squeak)	DIVISION	Replant in sun or shade 35–45 cm.	September/October	Edging plants or groups at the front of the border. Grow about 30 cm in height. They flower in March and April. Evergreen.
	SEED	Sow in pots in a propagator at 16–21°C. Move to a nursery bed when danger from frost is over.	February to May	
Berkheya (herbaceous types)	DIVISION	Replant in sunny spot 30–45 cm apart.	In spring	Thistle-like plants growing to about 90 cm to 1.2 m in height.
Boltonia (False Chamomile)	DIVISION	Replant in open ground 45 cm apart.	March	Heights 90 cm to 1.8 m. Bears aster-like flowers during summer. Rarely needs staking.
	SEED	Sow in a prepared bed outdoors, thin and plant out in autumn.	April	
Boykinia	DIVISION	Replant in woodland type soil 23–30 cm apart. Can also be raised from seed where available.	Autumn or spring	Plants most suited to the wild garden where they will grow 30–60 cm in height and flower in early summer.
Brunnera (B. macrophylla, syn. Anchusa myosotidiflora = Siberian Bugloss)	DIVISION	Replant in moist soil in sun or slight shade.	Autumn or spring	Height 30–45 cm. Plant out 23–30 cm apart. They flower in June. Border or wild garden plants.
	ROOT CUTTINGS	Root in pots in a cold frame.	During winter	
Buphthalmum (Yellow Ox-Eye; B. speciosum = Heart-Leaf Ox-Eye)	DIVISION	Replant in sunny spots 60–90 cm apart.	In spring	Robust border plants growing 60 cm to 1.2 m in height. B. speciosum makes a good waterside plant.
	SEED	Sow in a prepared bed outdoors.	April to June	

PLANT NAME	METHOD OF PROPAGATION	TREATMENT REQUIRED	TIME OF YEAR PROPAGATED	CULTURAL NOTES
Callirhoe (Poppy Mallow)	SOFTWOOD CUTTINGS	Root in a cold frame or propagator and plant out about 45 cm apart. Can also be raised from seed if desired.	April/May	Plants for dry well-drained position in sun. Grow up to 60 cm in height and flower in summer.
Campanula (Bellflower)	SEED	Sow in open ground or in cold frames.	April/May	Many HP species for the border and rock garden all flowering during summer. Heights for border types vary from 23 cm to 1.5 m.
	DIVISION	Replant according to the species' needs.	Autumn or spring	
	SOFTWOOD CUTTINGS	Root short cuttings in slight heat in propagator or frame.	March/April	
Cardamine (Bittercress; C. pratensis = Cuckoo Flower)	DIVISION	Replant in moist positions and slight shade 23–30 cm.	Autumn or spring	Plants for moist borders or the waterside where they grow to about 60 cm in height. Flower in late spring/early summer.
	LEAF CUTTINGS (C. pratensis)	Root leaf with stalk in a moist shady position outdoors.	June/July	
Carduncellus	DIVISION	Replant in sunny positions 30 cm apart.	Autumn or spring	Thistle-like plants growing to around 60 cm in height. Flower in June and July.
Carlina (Carline Thistle)	SEED	Sow outdoors where they will flower and thin to 45 cm apart.	March to May	Height 60 cm. Border or wild garden plants where they will self-seed unless flower heads are removed when dead.
Catananche (Cupid's Love Dart)	SEED	Sow in a prepared bed outdoors in drills.	March/April	Height 60 cm. Make good border plants and flowers can be cut and dried for winter use. They flower during summer.
	DIVISION	Replant in sunny spots 45 cm apart.	In spring	
	ROOT CUTTINGS	Root in pots in a cold frame.	During winter	
Celsia (C. arcturus = Bear's Tail Mullien)	SEED	Sow in slight heat and plant out 60 cm apart in June.	March to May	Rather tender plants of which C. bugulifolia is the hardiest. Others make good cold house subjects in pots and C. arcturus in summer bedding schemes.
	HALF-RIPE CUTTINGS	Root 8–10 cm cuttings in a frame or propagator.	July/August	
Centaurea (Knapweed)	SEED	Sow in a cold frame and move to nursery beds outdoors later.	April to June	Heights of perennial types are from 45 cm to 3 m most needing dividing every 3–4 years. They flower during summer.
	DIVISION	Replant in sun or slight shade planting from 30 cm to 45 cm apart.	Spring or autumn	
Cephalaria (C. gigantea = Giant Scabious)	SEED	Sow in a prepared bed outdoors and plant out in spring.	April	Tall plants for the back of the border. They flower in summer and are good for cutting.
	DIVISION	Replant 60 cm apart in sun or slight shade.	March/April	
Chelone (Turtle Head)	SEED	Sow in prepared bed outdoors.	April to June	Mid-border plants growing from 30 cm to 60 cm in height. They flower in late summer in sun or slight shade.
	DIVISION	Replant in well-drained soil 38 cm apart.	Spring or autumn	
	SOFTWOOD CUTTINGS	Root in a sand/peat compost in a cold frame.	In spring	
Chrysanthemum (C. maximum = Shasta Daisy; C. coccineum = Pyrethrum)	DIVISION	Replant in open ground 38 cm apart.	March or September	The two species beneath the title make good border plants or cut flower. They grow from 60 to 90 cm in height. C. cinerariifolium flowers are used to produce the insecticide 'pyrethrum'.
	SEED	Sow in a cold frame or earlier in slight heat and plant out in May.	March to May	

PLANT NAME	METHOD OF PROPAGATION	TREATMENT REQUIRED	TIME OF YEAR PROPAGATED	CULTURAL NOTES
Cimicifuga (Bugbane)	DIVISION	Replant in a moist soil 38–45 cm apart in slight shade.	In spring	Heights from 60 cm to 1.5 m depending on the species but all make good border plants. They flower in late summer.
	SEED	Sow in a prepared bed outdoors.	As soon as seed is ripe	
Claytonia (Spring Beauty)	DIVISION	Replant in moist peaty soil 23–30 cm apart.	In spring	Low growing succulent-like plants for the wild garden where they flower in spring. Can become a weed in borders.
	SEED	Sow in a prepared bed outdoors and plant out in spring.	As soon as seed is ripe	
Clematis (Traveller's Joy)	SEED	Sow in a cold frame or earlier in slight heat. Plant out late May.	March to May	Genus contains a number of herbaceous perennials for the border with species growing from 30 cm to 1.2 m in height. Flower in summer.
	DIVISION	Replant in lime-rich soil according to the species' requirements.	Autumn or spring	
Clintonia (C. uniflora = Queencup)	DIVISION (of rhizome)	Replant in a leafy soil 30 cm apart. Can also be raised from seed.	In spring	Flowering plants for shady positions. Grow from 15 to 45 cm in height.
Codonopsis (Bellwort)	SEED	Sow in a cold frame and grow on in nursery beds outdoors.	March/April	Mostly low-growing scrambling or medium-sized climbing plants. When used in a border they all need well staking.
	SOFTWOOD CUTTINGS	Root in slight heat and plant out when growing strongly.	In spring	
Commelina (C. coelestis = Blue Spiderwort)	DIVISION (of tuber)	Start into growth under glass, divide tubers, grow on and plant out in late May.	Early spring	Half-hardy perennials with tuberous roots. They can be stored like dahlias over winter in all but very mild areas where they can be left in the ground. Grow 45–60 cm tall.
	SEED	Sow in slight heat and grow on under glass. Harden off and plant out 30 cm apart in late May.	February/March	
Coreopsis (Tickseed)	SEED	Sow in prepared bed outdoors and plant out 38–45 cm apart in autumn.	April	Useful border plants or as cut flower. Most species grow 60–90 cm in height. They flower for long periods if dead flowers are removed.
	BASAL CUTTINGS	Root in a cold frame and overwinter there.	September/October	
Cortaderia (C. argentea = Pampas Grass)	SEED (surface sow)	Sow in a prepared bed outdoors or in slight heat indoors.	March to May	Growing to about 2 m in height, these plants are best grown as specimens. Division is best form of propagation as seed is variable.
	DIVISION	Divide and replant 90 cm apart in sheltered sunny spots.	In April	
Crambe (C. cordifolia = Sea Kale)	ROOT CUTTINGS	Root in pots in a cold frame. Plant out when established.	During winter or early spring	Another plant best as specimen as it grows 1–1.8 m in height and as much across. Takes 3 years to reach flowering size from seed.
	SEED	Sow in a prepared bed outdoors, thin, and plant out in autumn.	March to May	
Cynoglossum (Chinese Hound's Tongue)	DIVISION	Replant in sun or slight shade 30 cm apart.	March	Short-lived perennials growing 30–60 cm in height. C. amabile is best treated as an HB. They flower in early summer.
	SEED	Sow outdoors where they will flower and thin as above.	March/April	
Delphinium (Perennial Larkspur)	SEED	Sow in shallow drills outdoors. Plant out in September 45 cm apart.	April to June	Many strains and varieties mostly growing about 1.5 m in height but dwarf strains are available. Grow in well-manured soil in full sun. Can be used as cut flowers. Division unsatisfactory as plants resent the disturbance.
	BASAL HEELED CUTTINGS (named varieties)	Root 15 cm cuttings in pots in a cold frame.	March/April	
	DIVISION	Replant in growing positions.	March	

PLANT NAME	METHOD OF PROPAGATION	TREATMENT REQUIRED	TIME OF YEAR PROPAGATED	CULTURAL NOTES
Dicentra (D. spectabilis = Bleeding Heart/Lucy Locket/Dutchman's Breeches)	DIVISION	Replant where roots have a cool run.	Autumn or spring	Long established border plants growing about 45–60 cm in height. Roots are fragile so handle with care when dividing. Seed takes around 3 years to reach flowering size.
	ROOT CUTTINGS	Root in a sandy compost in cold frames.	January/February	
	SEED	Sow in cold frame or greenhouse and plant out 45 cm apart in spring.	As soon as ripe or in spring	
Dictamus (D. albus = Dittany/ Burning Bush)	DIVISION	Divide carefully and replant 38 cm apart.	In autumn	Border plant or cut flower growing 45–60 cm in height. Flowers and stem are rich in a volatile oil which, for amusement, can be ignited with a match on still warm days.
	SEED	Sow in beds outdoors and keep moist until germination occurs.	When seed is ripe or in spring	
Digitalis (Foxglove)	SEED (surface sow)	Sow on a prepared bed outdoors or in a cold frame. Plant out 30–45 cm apart in autumn.	May to July	Some species are HB and must be raised annually. All are well suited to a shady border. Heights are from 60 cm to 1.5 m.
	DIVISION (HP types)	Replant in flowering positions.	In spring	
Dispacus (D. fullonum = Fuller's Teasel)	SEED	Sow in shallow drills in a prepared bed outdoors. Plant out in sun or slight shade 45 cm apart in autumn.	May/June	A hardy biennial plant 1–1.8 m in height making good border or cut flower plant. Seed heads can be dried for floral arrangements.
Dodecatheon (Shooting Stars/ American Cowslip)	DIVISION	Replant in shady, moist positions 23 cm apart.	Autumn or spring	Border or wild garden plants growing 30–60 cm in height. They flower during May and June. D. meadia can be grown and housed in a cold greenhouse for early spring flowering.
	SEED	Sow in pots and place out to stratify then place in a cold frame in March.	December/January	
Doronicum (Leopard's Bane)	DIVISION	Divide every 2–3 years and replant in moist, cool soil 30–38 cm apart.	September	Among the first to flower in the year these plants grow 30–60 cm in height. Use either in the border or as cut flower.
	SEED	Sow in shallow drills outdoors.	April to June	
Dracocephalum (Dragon's Head)	DIVISION	Replant in semi-shade 30 cm apart.	In spring	The herbaceous types flower during summer, most growing 45–60 cm in height.
	SOFTWOOD CUTTINGS	Root in a cold frame and plant out when established.	April/May	
Echinacea (Purple Coneflower)	DIVISION	Divide carefully and replant 38–45 cm apart.	In spring	Height 60–75 cm. Prefer warm sunny positions in rich soil and little disturbance once established.
	SEED	Sow in a prepared bed outdoors.	April to June	
Echinops (Globe Thistle)	DIVISION	Replant in sun or shade 60 cm apart.	Spring or autumn	Best grown among other plants for foliage effects. There are species to flower from 60 cm to 1.8 m in height.
	ROOT CUTTINGS	Root in pots in a cold frame.	During winter	
	SEED	Sow in beds outdoors.	April to June	
Eomecon (Dawn Poppy/Snow Poppy)	DIVISION (of rhizome)	Replant in cool shady positions 45 cm apart.	In spring	Height 30 cm. Not very hardy in cold areas. Flowers in June.
Epimedium (Barrenwort/Bishop's Hat)	DIVISION	Replant in moist shady positions 30 cm apart.	After flowering	Mostly grown for its foliage and ground covering abilities. None grow more than 38 cm in height.
Eremurus (Foxtail Lily/Desert Candle/Giant Asphodel)	DIVISION	Divide every few years and replant 15–25 cm deep 45 cm apart.	October	Many flowering stems reach 1.5–1.8 m but foliage is low. Good mid-border plants. In cold areas roots are best lifted and stored like dahlias in winter.
	SEED	Sow in slight heat and grow on in cold frames.	March/April	

PLANT NAME	METHOD OF PROPAGATION	TREATMENT REQUIRED	TIME OF YEAR PROPAGATED	CULTURAL NOTES
Erigeron (Fleabane)	DIVISION	Replant in sun or slight shade 38 cm apart.	In spring	Many species and varieties most growing around 45 cm to 60 cm in height. Use in borders and as cut flower.
	SEED	Sow in drills in the open ground.	April to June	
Erodium (Heronsbill)	SEED	Sow in a cold frame then move to a nursery bed outdoors until autumn.	March/April	*E. manescavi* grows 30–60 cm in height and makes a good border plant, other species are more suited to the rock garden. Plant *E. manescavi* about 45 cm apart.
	ROOT CUTTINGS	Root and grow on in a cold frame. Plant out in a sunny well-drained position.	During winter	
Eryngium (Sea Holly)	ROOT CUTTINGS	Root in pots in a cold frame.	February/March	Striking foliage plants with species of all heights up to 1.8 m. Grow well in seaside localities. Flowers can be dried for winter use.
	SEED	Sow in pots and place out to stratify moving to a cold frame in March.	December/January	
Eupatorium (E. cannabinum = Hemp Agrimony; E. purpureum = Joe Pye Weed)	DIVISION	Replant in moist positions 50 cm apart.	In autumn or spring	Hardy perennials up to 1.5 m in height. Are best suited to the back of the border or wild garden. Some species are greenhouse plants.
	SEED	Sow in drills in a prepared bed outdoors. Plant out in autumn.	April to June	
Euphorbia (Milkweed/Spurge)	DIVISION (all except E. wulfenii)	Replant outdoors spacing according to species.	October or April	The herbaceous types provide a number of border subjects and some best grown as specimen plants. Their latex sap is poisonous so handle with care especially when using as cut flower.
	SEED	Sow in a prepared bed outdoors. Plant out in autumn.	May/June	
	SOFTWOOD CUTTINGS	Seal cut end with charcoal and root in a cold frame. Plant out 90 cm apart.	In spring	
Ferula (Giant Fennel)	SEED	Sow in drills outdoors and plant out in autumn 1.2 m apart.	May/June or as ripe seed is available	Tall plants up to 3.5 m in height. Best as specimen plants. Resent disturbance once established.
Filipendula (Dropwort/ Meadowsweet)	DIVISION	Replant in sun or slight shade 30–38 cm apart in moist positions.	In spring	Species with heights from 60 cm to 2.4 m. Need cool root runs in moist soils and a summer mulch. They flower during summer. Protect from frost in first few winters.
	SEED	Sow in pots or boxes in a cold frame and overwinter there.	In autumn	
Gaillardia (Blanket Flower)	DIVISION	Replant in sunny well-drained positions 30–38 cm apart.	In autumn	Perennial types make good cut flower or mid-border plants. They flower all through the summer months. Can also be raised from root cuttings.
	HALF-RIPE CUTTINGS	Root in a cold frame and overwinter there.	August/September	
	SEED	Sow in a cold frame and overwinter there.	May/June	
Galax (G. aphylla = Wand Flower)	DIVISION	Replant in open ground 30 cm apart.	In spring	An evergreen ground cover plant for moist shady borders. Height 30–60 cm.
Galega (Goat's Rue)	SEED	Sow in shallow drills in beds outdoors.	May/June	Growing 90 cm to 1.5 m this is a plant best at the back of the border as it is rather untidy in growth.
	DIVISION	Replant 60 cm apart.	Spring or autumn	
Gaura	SEED	Sow in sheltered spot outdoors and plant out 45–50 cm in autumn.	April/May	*G. lindheimeri* grows 90 cm to 1.2 m in height and makes a good border plant or raised early under glass an HHA.

PLANT NAME	METHOD OF PROPAGATION	TREATMENT REQUIRED	TIME OF YEAR PROPAGATED	CULTURAL NOTES
Gazania (Treasure Flower)	SEED	Sow in slight heat and plant out late May after hardening off.	March/April	Another perennial often treated as an HHA. They are rather tender but selected strains can be overwintered as cuttings. Plant out 30 cm apart.
	BASAL CUTTINGS	Root basal sideshoots in a warm propagator and keep frost-free through winter.	July/August	
Gentiana (G. asclepiadea = Willow Gentian)	SEED	Sow in a shady spot outdoors and plant out in spring 38 cm apart.	In autumn when seed is ripe	Genus contains a few herbaceous plants preferring moist soils in slight shade. G. asclepiadea grows 45 cm in height.
Geranium (Cranesbill)	SEED	Sow in drills in a prepared bed outdoors.	March to June	Many species for the border and the rock garden. Species to grow from 23 cm to 90 cm in height.
	DIVISION	Replant in growing positions 30–45 cm apart depending on species.	In spring or autumn	
Geum (Avens/Herb Bennet)	DIVISION	Divide every 3 years and replant in sun or slight shade 30–45 cm apart.	After flowering or early spring	Height 30–60 cm. Grow in a spot of their own in the border as their colours are very bright and likely to clash with other bright flowers.
	SEED	Sow in a prepared bed outdoors and plant out in autumn.	May/June	
Grass (HP) (Genera include: Avena, Bouteloua, Asperella, Festuca, Holcus, Luzula, Millium, Miscanthus, Pennisetum, Phalaris and Stipa)	SEED	Sow in a prepared bed outdoors in drills. Plant out in autumn.	March to June	Many genera and species suited to borders for foliage effect and/or flowers. A number such as Festuca ovina glauca make good edging plants planted about 25 cm apart.
	DIVISION	Divide and replant according to the needs of the species.	In spring	
Gypsophila (Chalk Plant)	SEED	Sow in slight heat and plant out in late May 60 cm apart in most cases. Plant in deep soil in sun.	February/March	G. paniculata and its cultivars are most used in the herbaceous border. They grow up to 90 cm in height but there are dwarf strains. Commercially double forms are wedge grafted under glass using 5 cm scions and 2-year-old single seedlings as the rootstock.
	BASAL HEEL CUTTINGS (named varieties)	Grow stock plants on in slight heat and root cuttings in a warm propagator or under mist.	In spring	
Helenium (Sneezeweed/Henbane)	DIVISION	Replant in sunny positions 45 cm apart.	Spring or autumn	Many varieties most growing 90 cm in height. Good border plants or cut flower. Flower in late summer.
	SEED	Sow in shallow drills outdoors. Plant out in late September/October.	April to June	
Helianthus (H. decapetalus = Perennial Sunflower)	DIVISION	Replant in sun 50 cm or so apart. Inclined to spread rapidly if not kept in check.	In spring	Tall plants up to 1.8 m in height for the back of the border or wild gardens. Flower in late summer.
Heliopsis (Orange Sunflower)	DIVISION	Replant in sun 45–50 cm apart.	In spring	Heights 90 cm but otherwise similar to above. Plant seedlings out in their final positions in autumn.
	SEED	Sow in shallow drills outdoors.	April to June	
Helleborus (H. niger = Christmas Rose; H. orientalis = Lenten Rose)	DIVISION	Take small divisions from young plants and plant out 45 cm apart.	October	Growing 30–60 cm in height these plants flower early in the year often in January. Grow in a deep soil in slightly shaded positions. Plant out in October or March.
	SEED (germination slow)	Sow in pots and place out to stratify then move to a cold frame in March.	December/January or as soon as seed is ripe	
Hemerocallis (Day Lily)	DIVISION	Replant in sun or shade 45 cm apart.	In spring	Easily grown plants up to 90 cm in height. There are many varieties making good border plants or cut flower. Seedlings flower in 2 years.
	SEED	Sow in a cold frame and plant out in final positions in autumn.	April to June	

PLANT NAME	METHOD OF PROPAGATION	TREATMENT REQUIRED	TIME OF YEAR PROPAGATED	CULTURAL NOTES
Hesperis *(Sweet Rocket/Damask Violet)*	SEED	Sow in shallow drills outdoors and plant out in autumn 30–45 cm apart.	May/June	Fragrant flowering plants 60–90 cm in height. Best grown as an HB though a good perennial for wild gardens.
	DIVISION	Divide every other year if grown in the border.	Autumn or spring	
Heuchera *(Alumroot/Coral Bells; Heucherella is treated in a similar manner)*	DIVISION	Replant in sun or shade 38 cm apart.	In spring	Low-growing plants for the front of borders. They flower through summer and are also good for cutting.
	SEED	Sow in shallow drills outdoors and plant out in autumn.	April to June	
Hieracium *(Hawkweed: H. aurantiacum = Devil's Paintbrush)*	DIVISION	Replant in well-drained positions 25 cm apart.	In spring	Height 30–45 cm. Pretty flowers for the front border. Inclined to spread by self-seeding if allowed to.
Hosta *(Plantain Lily)*	DIVISION	Replant in shady, moist positions 50 cm apart.	In spring	Mostly grown for their foliage though they do flower. There are a number of species and varieties all growing around 45–60 cm in height. Makes good ground cover.
	SEED	Sow in boxes in a shady spot outdoors and keep moist while germination occurs and for a month afterwards.	April to June	
Incarvillea *(Chinese Trumpet Flower)*	SEED	Sow in pots in a cold frame and plant out when established.	As soon as seed is ripe or in spring	Exotic looking plants growing 30–45 cm in height. Grow in sun and protect from frost and excessive wet in winter in colder areas.
	DIVISION	Divide carefully and replant just below soil surface 30–38 cm apart.	In spring as growth begins	
Inula *(Fleabane)*	SEED	Sow in shallow drills outdoors and plant out in autumn.	April to June	A number of species for moist soils in sunny situations for borders or wild gardens.
	DIVISION *	Replant according to the species.	October or March	
Iris *(rhizomatous types)*	DIVISION (of rhizome)	Lift and divide every 4 years or so, cutting rhizome into sections discarding centre of old plants. Replant with rhizome just showing above soil surface.	July or after flowering.	Many species and varieties for borders, cut flowers or pot plants. Plant rhizome-bearing types 15–25 cm apart depending on height of the type.
Isatis *(I. tinctoria = Woad)*	SEED	Sow in shallow drills outdoors and move to final positions in autumn 45 cm apart.	April to June	Though perennial, *I. glauca* and *I. tinctoria* are best treated as HB. They grow to between 60 cm and 1.2 m in height.
Kentranthus (Centranthus) *(Valerian/Pretty Betsy)*	SEED	Sow in a prepared bed outdoors and plant out 45 cm apart in autumn.	April to June	A good border plant especially for dry positions. Grows to 75 cm in height.
Kirengeshoma *(K. palmata = Yellow Wax Bells)*	DIVISION	Replant in moist soil and slight shade 50 cm apart. Can also be raised from seed when available.	March	Mid-border plant growing 75 cm in height. May need some protection from spring frosts.
Kniphofia *(syn. Tritoma = Red Hot Poker/Torch Lily)*	DIVISION	Replant in sunny well-drained positions 45–60 cm apart.	April	Many species and varieties from 45 cm to 1.8 m in height. Various flowering times through summer to autumn. Protect from winter wet and frost by tying foliage up above the crown.
	SEED	Sow in drills outdoors and plant out in September.	May/June	
K. caulescens and *K. northiae*	DIVISION (offsets: Irishman's cuttings)	Remove offsets when just rooted and grow on in a propagator until established.	April	
Lactuca *(L. bourgaei = Blue Lettuce)*	SEED	Sow in drills in a prepared bed outdoors. Plant out in autumn.	March to June	The lettuce family provide a number of plants for the border mostly tall and best given a back position.
	DIVISION	Replant in well-drained positions 60 cm apart.	In spring	

PLANT NAME	METHOD OF PROPAGATION	TREATMENT REQUIRED	TIME OF YEAR PROPAGATED	CULTURAL NOTES
Lamium *(Dead Nettle)*	ROOT CUTTINGS	Root in a cold frame and plant out 30 cm apart.	February/March	The cultivated types make good edging in sun or slight shade. Heights vary from 20 to 60 cm.
	DIVISION	Replant in open ground.	March	
	SEED	Sow in a prepared bed outdoors.	March	
Lathyrus *(Everlasting Pea)*	DIVISION	Replant in open ground; climbers about 60 cm apart, others 38 cm apart.	March	Genus contains a number of climbers up to 1.8 m in height and a few border plants of 30–60 cm. Grow in sheltered positions.
	SEED	Sow where they will flower or in nursery beds.	March to May	
Lavatera *(L. olbia = Tree Mallow)* *(sow HA types where they will flower March to June)*	SEED	Sow in a prepared bed outdoors and plant out 60 cm apart in autumn.	March to June	Shrubby plants, the perennial types being 1.5–1.8 m in height. They grow best in good soil and full sun. Overwinter autumn struck cuttings in a cold frame.
	SOFTWOOD CUTTINGS	Root in slight heat and grow on in a nursery bed outdoors.	As soon as growth begins in spring	
	HALF-RIPE HEEL CUTTINGS	Root in a warm propagator.	September	
Liatris *(Kansas Feather/Gay Feather)*	SEED	Sow in slight heat and grow on in a nursery bed outdoors.	March/April	Useful border plants and as cut flower with species from 60 cm to 1.5 m in height.
	DIVISION	Replant in sun 30 cm apart.	March	
Ligularia *(Ragwort. L. clivorum, syn. Senecio clivorum = Golden Groundsel)*	DIVISION	Replant in moist soil in sun 60 cm apart.	March	Mostly grown for foliage effect though they also flower. They are better in the wilder parts of the garden where they grow 90 cm to 1.5 m in height.
	SEED	Sow in drills in a prepared bed outdoors. Plant out in September/October.	April to June	
Limonium *(syn. Statice: HP types = Sea Lavender)*	SEED	Sow in drills in a prepared bed outdoors.	April	All species are useful for borders or as cut flowers which can be dried as 'everlastings'. *L. sinuatum* and *L. suworowii* are HHAs. HP types grow about 60 cm in height.
	DIVISION	Divide carefully and replant 50 cm apart.	In spring	
L. latifolium	ROOT CUTTINGS	Root in a cold frame and plant out when established.	February/March	
Linaria *(Toadflax; HP types)*	DIVISION	Replant in sun or slight shade 30–45 cm apart depending on species.	In spring or autumn	Perennial types mostly grow 60–90 cm in height. They flower all through summer.
Linum *(Flax; HP types)*	SEED	Sow in a prepared bed outdoors.	April	Like *Linaria* these plants flower over a long period and grow 30–60 cm in height. They need a well-drained spot and plenty of sunshine.
	DIVISION	Divide carefully and replant 38 cm apart.	In spring	
	HALF-RIPE CUTTINGS	Root in a cold frame or propagator.	July	
Lobelia *(Perennial types: L. cardinalis = Cardinal Flower)*	SEED (surface sow)	Sow in slight heat and harden off before planting out in late May/early June.	February/March	Beautiful plants up to 90 cm in height. They are rather tender and should be lifted to overwinter in frost-free conditions in all but very mild areas. Grow in sun or slight shade.
	DIVISION	Lift or divide when replanting in spring and plant 30 cm apart in moist soil.	May	
Lupinus *(Lupin)*	BASAL HEEL CUTTINGS	Root 8–10 cm cuttings in slight heat. Pot on and plant out 45 cm apart in May.	In spring as soon as growth is large enough	Plants best grown in groups for effect. They flower around 75–90 cm in height. Grow in sun or slight shade and plant firmly. Seed will not flower true.
	SEED	Sow in slight heat or in the open in May and June.	February/March	

PLANT NAME	METHOD OF PROPAGATION	TREATMENT REQUIRED	TIME OF YEAR PROPAGATED	CULTURAL NOTES
Lychnis (Campion: L. chalcedonica = Maltese Cross)	SEED	Sow in drills in a prepared bed outdoors.	April to June	L. chalcedonica grows well at the back of the border where it grows up to 90 cm. Strains of some British wild species are also cultivated.
	DIVISION	Replant in sun 30–38 cm apart depending on the species.	In spring	
Lysimachia (Loosestrife)	SEED	Sow in drills in a prepared bed outdoors. Plant out in autumn.	April to June	Vigorous plants for sun or slight shade where they grow 60–90 cm in height. Some species make good waterside plants.
	DIVISION	Replant 38–50 cm apart in moist soils.	Autumn or spring	
Macleaya (Plume Poppy)	SEED	Sow in drills in a prepared bed outdoors.	April to June	Large plants up to 2.4 m in height ideal for borders or as specimen groups. Plant well-rooted cuttings out in autumn or leave in the cold frame for the winter.
	STEM CUTTINGS	Make cuttings from sideshoots and root in a shady cold frame. Plant out 60 cm apart.	July/August	
Malva (M. moschata = Musk Mallow)	SEED	Sow in slight heat under glass and plant out 45 cm apart when established.	March	These plants grow some 60–90 cm in height and flower all through summer. M. crispa is an HA foliage plant growing 1.8 m in height.
	SOFTWOOD CUTTINGS	Make 8–10 cm cuttings and root in a warm propagator.	In spring	
Meconopsis (HP types: M. betonicifolia = Himalayan Blue Poppy; M. cambrica = Welsh Poppy)	SEED	Sow thinly in pots in a shaded cold frame or in very slight warmth. Plant out in late spring 30 cm apart.	As soon as seed is ripe or in spring	A number of species well suited to damp, shady woodland type soils. M. betonicifolia is best treated as a biennial. Heights are from 30 cm to 1.5 m.
Mertensia (M. virginica = Virginian Cowslip)	SEED	Sow in a prepared bed outdoors.	As soon as seed is ripe	Plants for semi-shade growing up to 60 cm in height. They flower early in the year.
	DIVISION	Replant 30 cm apart in moist soils.	October or March	
Morina (Himalayan Whorlflower)	DIVISION	Replant in sheltered positions 30 cm apart.	In spring	Thistle-like plants growing 60–90 cm in height. Transplant seedlings to permanent positions as soon as large enough to handle.
	SEED	Sow in a prepared bed outdoors and shade until established.	As soon as seed is ripe	
Nepeta (N. faassenii, syn. N. mussinii = Catmint)	SOFTWOOD CUTTINGS	Root in a cold frame and overwinter there.	August/September	Useful border or edging plants up to 45 cm in height. Cut down to ground after flowering to produce cuttings for late summer.
	DIVISION	Replant in semi-shade in well-drained soil 38 cm apart.	March	
Oenothera (Evening Primrose: O. fruticosa = Sundrops)	SEED	Sow in drills in a prepared bed outdoors.	June/July	A very varied genus in heights and habits. O. missouriensis is a trailing plant 23 cm in height, but most border types grow to around 60 cm.
	DIVISION (perennial species)	Replant in sunny well-drained positions 30–45 cm apart.	March/April	
	SOFTWOOD CUTTINGS	Root in a shady cold frame. Plant out in autumn.	May	
Omphalodes (Navelwort)	DIVISION	Replant in slight shade 25–30 cm apart. Can also be raised from seed.	September/October	Low-growing plants up to 30 cm in height ideal for the front of shady borders.
Onopordum (O. acanthium = Cotton Thistle)	SEED	Sow in a prepared bed outdoors and plant out in autumn 60–90 cm apart.	April to June	Growing 1.2–1.5 m in height this thistle-like plant is ideal as a specimen plant especially in exposed situations.
Pachysandra	DIVISION	Replant in shade 25–30 cm apart.	Autumn or spring	Evergreen ground cover plants up to 30 cm in height. Grow well in deep shade beneath trees.

PLANT NAME	METHOD OF PROPAGATION	TREATMENT REQUIRED	TIME OF YEAR PROPAGATED	CULTURAL NOTES
Paeonia (Paeony or Peony)	DIVISION (of named varieties)	Divide carefully and replant in sun or slight shade 60 cm apart.	September/October	Most herbaceous species grow between 60 cm and 90 cm in height. Plant in groups for effect. Seedlings will only come true for species and take up to 4 years to reach flowering size.
	SEED	Sow in a cold frame and plant out in September or in spring.	As soon as ripe or in spring	
Papaver (Poppy)	SEED	Sow in drills in a prepared bed outdoors. Plant out in autumn 38–45 cm apart.	April to June	Spectacular flowers on plants 60–90 cm in height. Grow in well-drained sunny positions in groups for effect. Best in a mid-border position.
	ROOT CUTTINGS (named types)	Root in pots in a cold frame planting out when established.	February/March	
Patrinia	DIVISION	Replant in moist shady position 30 cm apart. Can also be raised from seed.	Autumn or spring	Flowering in July at about 30 cm in height these are ideal front border subjects.
Penstemon (Beard Tongue)	SEED	Sow in slight heat and plant out in late May/early June 45 cm apart.	February/March	Rather tender plants best overwintered as cuttings in cold areas. Most species grow about 60 cm in height and are best grown in sun.
	BASAL CUTTINGS	Root in a cold frame and overwinter there.	September	
Perovskia (Russian Sage: occasionally spelt Perowskia)	HALF-RIPE CUTTINGS	Root in a cold frame and plant out in sunny positions 60 cm apart in autumn or spring.	July	Deciduous shrub-like plants growing up to 1.5 m in height. Cut back to 45 cm annually.
Phlox (HP types)	ROOT CUTTINGS	Root in a cold frame and overwinter there. Plant out in spring in a nursery bed to grow on.	August/September	Best grown in clumps in mid-border positions where most grow 75–90 cm in height. Grow in sun or slight shade planting out 45 cm apart. They flower in July and August.
	SEED (slow germination)	Sow in slight heat and grow on in warmth until the following spring.	Autumn as soon as seed is ripe.	
Phormium (New Zealand Flax)	SEED (does not come true)	Sow in drills in a prepared bed outdoors and plant out 75 cm apart in autumn.	April/May	Sword-leaved foliage plants ideal as specimens as they grow upwards of 1.8 m in height in moist soils and sunny situations.
	DIVISION	Divide to retain a fan of leaves throwing away any with flowers.	Spring, before growth begins	
Physalis (Cape Gooseberry/ Chinese Lantern)	SEED	Sow in drills in a prepared bed outdoors.	April to June	Plants up to 45 cm in height mostly grown for their ornamental fruits which are used in floral decorations when dry.
	DIVISION	Replant in sheltered positions in sun or slight shade 38 cm apart.	April	
Physostegia (Obedient Plant)	DIVISION	Replant in sun or slight shade 45 cm apart.	In spring	Plants growing from 45 cm to 90 cm in height peculiar in that its flowers will remain where placed on the flower stalk.
	SOFTWOOD CUTTINGS	Root in a propagator or cold frame and plant out when established.	In spring	
Phytolacca (Pokeberry)	SEED	Sow in drills in a prepared bed outdoors.	April to June	*P. americana* is the tallest of these plants which are best at the back of the border where it can grow up to 3 m in height.
	DIVISION (poisonous roots and berries)	Replant 75 cm apart in slight shade and moist soil.	In spring	
Platycodon (Chinese Bell Flower/Balloon Flower)	SEED	Sow in boxes in a cold frame and grow on in a nursery bed outdoors.	March to May	Ideal for front border positions as they grow 30–45 cm in height. Plant in well-drained sunny positions.
	DIVISION	Divide carefully and replant 30 cm apart.	In spring	
Podophyllum (May Apple)	DIVISION	Replant 38–45 cm apart in moist soils and shady positions.	In autumn	Fine foliage plants growing up to 60 cm in height. Leaves and roots are poisonous but fruits are edible.
	SEED	Sow in a cold frame and overwinter there.	As soon as seed is ripe	

PLANT NAME	METHOD OF PROPAGATION	TREATMENT REQUIRED	TIME OF YEAR PROPAGATED	CULTURAL NOTES
Polemonium *(Jacob's Ladder/Greek Valerian)*	SEED	Sow in drills in a prepared bed outdoors.	March to May	Genus contains a number of species for border work from 15 cm to 90 cm in height.
	DIVISION	Replant in sun or slight shade 25–38 cm apart.	Autumn or spring	
Polyanthus *(Primula veris × P. elatior × P. vulgaris hybrids)*	SEED (surface sow)	Sow in a shady propagator at 13–16°C and keep moist and growing in cool conditions.	June	Ideal plants for spring bedding or permanent place at the front of the border. Many spectacular varieties. Best grown in moist soils and slight shade.
	DIVISION	Replant 30 cm apart.	After flowering	
Polygonum *(Knotweed; HP types)*	DIVISION	Replant according to the species in sun or slight shade. Some species can also be raised from seed when available.	February/March	A wide range of species from the 15–30 cm *P. vacciniifolium* for front of the border to *P. sachalinese* growing 2.5–3.5 m high.
Potentilla *(Cinquefoil; HP types)*	DIVISION	Replant in sun or slight shade in dry positions 30–45 cm apart.	Spring or autumn	There are a number of border species from 30 to 60 cm in height. They are best grown in groups for effect.
	SEED	Sow in drills in a prepared bed outdoors. Plant out in autumn.	April/May	
Pulmonaria *(Lungwort)*	DIVISION	Replant in sun or slight shade 30–45 cm apart depending on the species.	Spring or autumn	Flowering in March/April at around 30–45 cm high these plants rapidly self-seed unless dead-headed.
Ranunculus *(Buttercup: R. acris = Bachelor's Buttons)*	SEED	Sow in drills in a prepared bed outdoors.	April to June	The border types are best grown in sunny positions in fairly moist soil. They grow up to 60 cm high.
	DIVISION	Replant 30–38 cm apart depending on the species.	In autumn	
Rodgersia	SEED	Sow in drills in a prepared bed outdoors.	April to June	Ornamental foliage plants growing 90 cm to 1.2 m in height. Grow in sheltered parts of the garden.
	DIVISION (of rhizome)	Replant in moist soils in sun or slight shade 50 cm apart.	In spring	
Romneya *(California Tree Poppy)*	SEED	Sow singly in small pots in slight heat and plant out without disturbance 1.2–1.5 m apart. Can also be divided in spring if done with great care.	March to May	Splendid but somewhat difficult border plants growing up to 1.8 m in height. Requires well-drained sunny positions and winter protection in cold areas.
Roscoea	SEED	Sow in slight heat and grow on in cold frame.	March to May	Orchid-like flowers on plants growing 30 cm in height. Ideal for front of the border or rock garden.
	DIVISION (of rhizome)	Replant 10–13 cm deep, 25 cm apart in sunny positions.	In spring	
Rudbeckia *(Coneflower: HP types)*	DIVISION	Replant in sunny positions 38–45 cm apart.	April	Border types make good cut flower and come in a variety of heights from 60 cm to 1.8 m. They grow best on heavy soils.
	SEED	Sow in drills in a prepared bed outdoors. Plant out in autumn.	April to June	
Saponaria *(S. officinalis = Bouncing Bet/Soapwort)*	SEED	Sow in a prepared bed outdoors and plant out in autumn 38 cm apart.	April to June	*S. officinalis* grows to 90 cm in height, but other perennial species are much shorter and are best in the rock garden. They are best in acid soils.
	SOFTWOOD CUTTINGS	Root short cuttings in a cold frame and plant out in sunny well-drained positions.	In spring	

PLANT NAME	METHOD OF PROPAGATION	TREATMENT REQUIRED	TIME OF YEAR PROPAGATED	CULTURAL NOTES
Scabiosa *(Scabious/Pincushion Flower)*	DIVISION	Replant 45 cm apart in moist soils in sunny positions.	In spring	Ideal cut flower or border plant most growing about 60 cm in height. They grow best on calcareous soils. Seedlings will produce plants in various shades.
	BASAL HEEL CUTTINGS	Root in a cold frame or greenhouse. Plant out when established.	As soon as growth begins in spring	
	SEED	Sow in drills in a prepared bed outdoors.	April to June	
Sedum *(Stonecrop; HP types; S. spectabile = Ice Plant)*	DIVISION	Replant 30–45 cm apart depending on the species in sunny positions. Can also be raised from seed where available.	March/April	A large genus with many types suitable for the rock garden. Border types grow up to 45 cm in height.
Senecio *(Groundsel/Ragwort)*	SEED	Sow in slight heat and plant out in May.	March/April	A large genus, most cultivated forms being grown for their grey foliage more usually termed 'silver'. *S. cineraria* is used in summer bedding as a HHA. Most are easily cultivated plants.
	SOFTWOOD CUTTINGS	Root in a cold frame or greenhouse and overwinter there if necessary.	Spring or autumn	
	DIVISION	Divide and replant according to the species.	April/May	
Sidalcea *(Prairie or Greek Mallow)*	SEED	Sow in a cold frame and overwinter there. Plant out in late spring 38 cm apart.	As soon as ripe or in spring	Long flowering plants growing from 75 cm to 1.2 m in height. Grow in sunny positions and cut well back after flowering.
	DIVISION	Replant as above.	In spring or autumn	
Silphium *(S. laciniatum = Compass Plant/Resinweed)*	DIVISION	Replant 60 cm apart. Can also be increased from seed when available.	In spring	Tall plants up to 1.8 m in height best grown in groups at the back of the border.
Sisyrinchium *(Satin Flower/Blue-eyed Grass)*	SEED	Sow in a cold frame and grow on there until planting out August or September.	March/April	Genus contains a number of species for rock garden and border, *S. striatum* being best for this purpose growing to 45 cm in height.
	DIVISION	Replant in sunny positions 45 cm apart.	In spring	
Smilacina *(S. racemosa = False Spikenard; S. stellata = Star-flowered Lily-of-the-Valley)*	SEED	Sow in a shady cold frame and grow on there until planting out 38 cm apart in autumn.	March to May	Plants for a moist shady border where they grow around 60 cm in height. They flower in May and June.
Solidago *(Golden Rod/Aaron's Rod)*	DIVISION	Replant in sunny positions 38–45 cm apart. Can also be raised from seed when available.	In spring	Border plants and cut flower growing from 60 cm to 1.2 m in height depending on the species and variety.
Stachys *(Wound-wort; S. lanata = Lamb's Ears)*	DIVISION	Replant in poor soils in sun or slight shade 30–38 cm apart.	March/April	Edging or front border plants with silver leaves which flower in summer. Height 30–60 cm.
	SEED	Sow in drills in a prepared bed outdoors.	April to June	
Stokesia *(Stoke's Aster)*	DIVISION	Replant in sunny positions 38 cm apart.	March/April	Autumn-flowering border or cut flower plants up to 45 cm in height.
Stylophorum *(Celandine Poppy)*	DIVISION	Replant in slight shade 30 cm apart.	In spring	A border or rockery plant growing 30 cm in height. They flower in April and May.
	SEED	Sow in drills in a prepared bed outdoors.	April/May	

PLANT NAME	METHOD OF PROPAGATION	TREATMENT REQUIRED	TIME OF YEAR PROPAGATED	CULTURAL NOTES
Tellima	DIVISION	Replant in sun or slight shade 38 cm apart.	In spring or autumn	*T. grandiflora* is an evergreen ideal for front border positions. The flower stems rise above the leaves to 60 cm.
	SEED	Sow in a cold frame and grow on in a nursery bed outdoors.	March to May	
Thalictrum *(Meadowrue)*	DIVISION	Replant in sun or slight shade 38–45 cm apart.	March/April	Plants for the back of the border mostly growing 90 cm in height but often more. *T. dipterocarpum* is propagated from offsets.
	SEED	Sow in drills in a prepared bed outdoors. Plant out in autumn.	April to June	
Thermopsis	SEED	Sow in a cold frame or prepared bed outdoors. Plant out in autumn 45 cm apart.	March to May	Lupin-like plants growing 60–90 cm in height and flowering in summer. Division of older plants is not recommended.
Tiarella *(Foam Flower)*	SEED	Sow in a cold frame and grow on in shady beds outdoors.	March to May	Ideal plants for moist peaty soil and shady positions. The flowers rise above the foliage to 30–60 cm in height.
	DIVISION	Replant 30 cm apart.	In autumn	
Tradescantia *(Spiderwort; T. virginiana = Trinity Flower)*	SEED	Sow in a cold frame or prepared bed outdoors.	March to May	*T. virginiana* and its varieties are the usual HP types found in borders where they grow around 45 cm in height.
	DIVISION	Replant in sun or slight shade 38 cm apart.	Early spring	
Tricyrtis *(Toad Lily)*	SEED	Sow in slight heat and grow on in a cold frame, planting out when frost has gone.	March/April	Rather tender plants growing best in a peaty soil in slight shade. They grow up to 60 cm in height. Give some frost cover over winter in cold areas.
	DIVISION	Remove offsets and replant 30 cm apart.	In spring	
Trollius *(Globe Flower)*	DIVISION	Replant in moist soils 38 cm apart in sun or very slight shade.	September	Buttercup-like flowers appear in May and June on plants growing between 30 and 90 cm in height depending on species and variety. Best in large groups in the border.
	SEED (slow germination)	Sow in a cold frame or a prepared bed outdoors. Plant out when large enough.	As soon as ripe or in spring	
Uvularia *(Bellwort)*	DIVISION (of rhizome)	Replant in moist peaty soil 15 cm apart.	In spring	Good for the front of moist and shady borders as they grow up to 45 cm in height. Flowers appear in May and June.
Veratrum *(False Hellebore)*	SEED	Sow in a moist shady prepared bed outdoors.	March to June	A few border species mostly growing to around 90 cm in height. They flower during summer.
	DIVISION (rhizomes and roots are poisonous)	Replant in rich, moist soils in light shade planting 45 cm apart.	In spring	
Veronica *(Speedwell; HP types)*	DIVISION	Replant in sunny positions 30–38 cm apart.	Early autumn or spring	Border types mostly grow 30 to 60 cm in height. They all flower over a long period during summer.
	SEED	Sow in drills in a prepared bed outdoors.	April/May	
Viola	BASAL CUTTINGS	Make cuttings 5 cm long and root in a cold frame and plant out early cuttings in autumn and later ones in spring spacing according to type.	June/July or August/September	A large genus containing violas, violets and pansies. Most violas are treated as perennials, pansies as biennials, but all can be raised from cuttings. Some border violas such as *V. cornuta* can also be divided.
	SEED	Sow in slight heat and grow on in cold frames.	June or September	

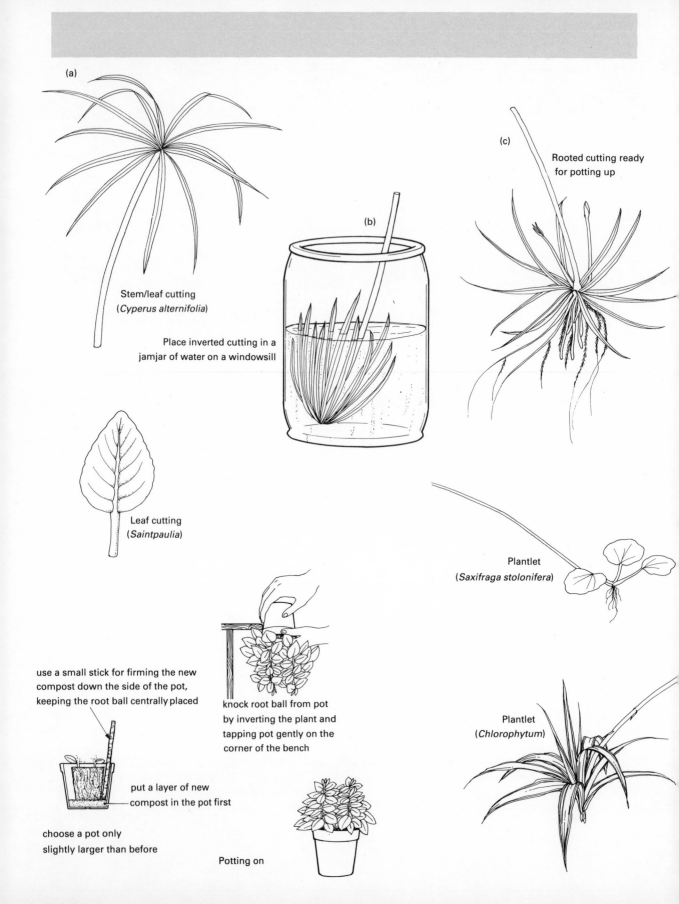

(a)

Stem/leaf cutting
(*Cyperus alternifolia*)

(b)

Place inverted cutting in a
jamjar of water on a windowsill

(c)

Rooted cutting ready
for potting up

Leaf cutting
(*Saintpaulia*)

Plantlet
(*Saxifraga stolonifera*)

use a small stick for firming the new
compost down the side of the pot,
keeping the root ball centrally placed

knock root ball from pot
by inverting the plant and
tapping pot gently on the
corner of the bench

Plantlet
(*Chlorophytum*)

put a layer of new
compost in the pot first

choose a pot only
slightly larger than before

Potting on

17 House and Greenhouse Plants

There is no better way to learn about plants than to live with them, and with a great number we can do just that by cultivating them in our own homes. A wide range are suitable for house culture, and still more can be added if a heated greenhouse is available to raise others as temporary visitors during their flowering season.

When it comes to propagation the same applies, for although such types as the Spider Plant (*Chlorophytum*) and Busy Lizzie (*Impatiens*) are easily propagated on the kitchen windowsill, many others are more exacting in their requirements, making a propagator or similar heated frame essential. Plants raised from seed will invariably benefit from a good start in an appropriately heated greenhouse, but in all cases remember to shade all seedlings and young plants from direct sunlight, especially as spring and summer progress.

For potting on the majority of plants for the house I prefer to use a peat-based compost, though a compost of two-thirds (by volume) John Innes Potting compost plus one-third moss peat is also suitable. Using the latter, the strength of the fertiliser content should in most cases be determined by the size of the plant and container, starting with No. 1 for the first potting, moving through 2 and only reaching 3 when the plants reach 23–25 cm pot proportions. For greenhouse potting the John Innes compost should be used without the addition of the peat. In the majority of cases do not overpot and go easy on the water, letting the compost become almost dry before watering again.

Greenhouses come in all shapes and sizes, but of most concern to us here are the temperatures which can be maintained in them, especially during winter. In the cultural notes contained in this chapter I have tried to indicate the minimum winter temperature required for each subject and it is to this minimum that any temperature refers. These temperatures apply both to house or greenhouse and it can also be taken that any plant suitable for household cultivation will do as well, if not better, in a greenhouse appropriate to its type.

As a further guide, the definition of greenhouse temperatures in this book is as follows. Cold house = no artificial heat; cool house = frost-free to 7 °C; warm house = 10–13 °C; and stove house = not less than 16 °C and preferably more.

PLANT NAME	METHOD OF PROPAGATION	TREATMENT REQUIRED	TIME OF YEAR PROPAGATED	CULTURAL NOTES
Abutilon (Flowering Maple/Indian Mallow)	SEED	Sow singly in small pots in a temperature of 18–21°C.	February/March	Small to medium-sized shrubs grown for foliage and flowers. Greenhouse or summer outdoors. Mostly 10°C though a few are almost hardy.
	HARDWOOD CUTTINGS	Make 5 cm cuttings and root in sand/peat in a propagator at 18–21°C.	September or March	
Acacia (Wattle; A. dealbata = Mimosa)	SEED	Sow in a temperature of 18°C.	March	Rather tender evergreen flowering shrubs best in a frost-free greenhouse but also outdoors in mild areas.
	HALF-RIPE CUTTINGS	Make 5 cm cuttings and root in a warm propagator.	July to September	
Acalypha (Copper Leaf)	SOFTWOOD CUTTINGS	Root in a propagator at 24–27°C.	March	Evergreen foliage plant for a stove house. 16–18°C.
Adenanthera (A. pavonina = Peacock Flower Fence/Red Sandalwood Tree)	SEED (warm water soak 24 hours before sowing)	Sow in a propagator at 21–24°C.	February to May	A. pavonina is an evergreen shrub up to 1.5 m in height. Stove house 16°C. Seeds are used to make necklaces.
	HALF-RIPE CUTTINGS	Root in sand in a propagator at 21–24°C.	During summer	
Aechmea (Urn Plant)	SEED	Sow in a propagator at 27–29°C.	March to May	Bromeliad houseplant. Grow in good light and keep central 'vase' filled with rainwater. Grow in small pots and do not overwater. 13–16°C.
	DIVISION (offsets/suckers)	Detach large offsets at their base, pot in small pots and grow on in a propagator at 21–27°C.	May/June	
Aeschynanthus (syn. Trichosporum)	HALF-RIPE or SOFTWOOD CUTTINGS	Root 5–8 cm cuttings in sand/peat in a propagator at 21–24°C.	April/May	Epiphytic trailing plants for a warm greenhouse. 13°C. Grow in shade. Pot into a compost of fibrous peat, sphagnum moss, and a little charcoal.
Aglaonema (A. commutatum = Chinese Evergreen; A. simplex = Malayan Sword)	SEED	Sow in a propagator at 27–29°C.	In spring	Foliage houseplant for shady position or a warm moist greenhouse. 13°C. Needs a steady temperature.
	HALF-RIPE CUTTINGS	Root in sand/peat in a propagator at 24–27°C.	During summer	
	DIVISION	Repot in small pots.	March	
Albizzia (Pink Siris Tree)	SEED	Sow in a propagator at 16–18°C.	March to June	Deciduous shrubs and trees for a cool greenhouse or outdoors in very mild areas.
Alocasia	DIVISION (of rhizome)	Repot in a peat-based compost with rhizome on surface. Establish in a warm propagator.	March/April	Medium-sized stove house foliage plants. 18°C.
Alpinia (Indian Shell Flower)	DIVISION (of rhizome)	Replant divisions in small pots to grow on.	March/April	Large perennials up to 3 m in height for warm greenhouses. 13°C.
Amasonia	STEM CUTTINGS	Root 8–10 cm cuttings in a propagator at 21°C.	In autumn	Evergreen flowering sub-shrubs 60–90 cm in height for a warm greenhouse. 13–16°C.
	ROOT CUTTINGS	Root in pots in a propagator at 21°C.	March/April	
Ananas (A. comosus = Pineapple)	DIVISION (crowns of plant or offsets)	Slice off a green crown horizontally with a thin layer of fruit attached. Press into a peat compost with a surface layer of sand. Root in a propagator at 24–27°C.	In spring	Pineapples can be grown as houseplants in full sun and will eventually form the well-known fruit. Offsets will do even better.
Anthurium (A. scherzerianum = Flamingo Flower)	SEED	Sow in well-drained pots in a temperature of 24–27°C.	February to April	Evergreen flowering houseplants for moist shady positions. 16°C.
	DIVISION	Replant with crowns above compost surface.	January	

PLANT NAME	METHOD OF PROPAGATION	TREATMENT REQUIRED	TIME OF YEAR PROPAGATED	CULTURAL NOTES
Antigonon (A. leptopus = Coral Vine)	SEED	Sow in a temperature of 21–24°C. Grow on in the greenhouse border.	In spring	Climbing flowering perennials for warm greenhouses where they need much heat and light.
Aphelandra (A. squarrosa = Zebra Plant)	STEM CUTTINGS	Root 10–15 cm sideshoots in a propagator at 24°C.	April	Evergreen shrubby houseplants requiring well lit positions and a rich compost. 13–16°C.
Arachis (A. hypogaea = Peanut/Monkeynut	SEED	Sow singly just below soil surface in large pots in a temperature of 22°C.	April/May	Greenhouse annual 30–60 cm in height. Will form peanuts after flowering and bury them to mature.
Araucaria (A. excelsa = Norfolk Island Pine)	SEED	Sow 2.5 cm deep in a propagator at 18°C.	March/April	A. excelsa is an evergreen coniferous tree making a good houseplant in sun or slight shade. Just keep frost-free. Some species are hardy.
	TIP CUTTINGS	Root in a peat-based compost in a warm propagator.	September	
Araujia (A. sericifera = Cruel Plant)	SEED	Sow in a temperature of 16–18°C.	March to June	Vigorous climbing plants for a cool greenhouse or outdoors in very mild areas.
	HARDWOOD CUTTINGS	Root in sand/peat in slight heat.	October/November	
Arcliontophoenix (A. cunninghamiana = Illawarra Palm)	SEED (warm water soak 48 hours before sowing)	Sow in a propagator at 21–24°C. Grow on in small pots at first.	In spring	Greenhouse palm requiring a 10°C (50°F) minimum in winter. Water well in summer.
Ardisia (Spear Flower)	SEED	Sow in a temperature of 21–24°C.	March/April	Cool greenhouse shrubs grown for their flowers and red berries. A. crispa and A. japonica make good houseplants.
	HALF-RIPE CUTTINGS	Root 5 cm cuttings in a warm propagator.	April/May	
Aristea	SEED	Sow in a temperature of 21–24°C.	March to May	Greenhouse flowering perennials 45–60 cm in height. 10°C.
Aristolochia (A. elegans = Calico Flower)	SEED	Sow in a temperature of 16–18°C.	March to May	Climbing shrubs with fragrant flowers for cool greenhouses. 7–10°C. Grow in the greenhouse border.
	STEM CUTTINGS	Root in a warm propagator.	June/July	
Asclepias (A. currasavinca = Blood flower)	SEED	Sow in a temperature of 18–21°C.	March to May	The greenhouse types are best in a stove house. 16°C. Nip outside tips to induce bushy growth.
	HALF-RIPE CUTTINGS	Root in a warm propagator. Start off in small pots.	In spring	
Asparagus (Asparagus Fern)	SEED	Sow in a temperature of 21–24°C.	March to May	Foliage houseplants for shade or semi-shade. 7–10°C in winter.
	DIVISION	Repot in small pots in a peat-based compost.	March	
Aspidistra (Cast Iron Plant/Parlour Palm)	DIVISION	Repot into small pots. Very slow growing.	In spring	Ideal foliage houseplant for shade or semi-shade. 7°C.
Azalea (A. indica, syn. Rhododendron simsii = Japanese Azalea)	HALF-RIPE CUTTINGS	Root 8 cm cuttings in peat/sand in a warm propagator. Pot on in a lime-free compost and 13 cm pot for flowering.	June/July	Evergreen shrubs used as houseplants in winter being placed in a shady spot outdoors in summer. Grow cool but frost-free.
Bauhinia (B. variegata = Ebonywood)	SEED	Sow in a temperature of 21–24°C.	March to May	Shrubs and small trees some such as B. galpinii are climbing and are suitable for cool greenhouses but others need more heat.
	HALF-RIPE CUTTINGS	Root in a propagator at 21–24°C. Use a sand/peat compost.	In spring	
Beaufortia	SEED	Sow in a temperature of 21–24°C.	March to May	Small flowering shrub up to 60 cm in height. Cool greenhouse. 7–10°C.
	HALF-RIPE CUTTINGS	Root in slight heat and grow on in a peat-based compost.	July/August	

PLANT NAME	METHOD OF PROPAGATION	TREATMENT REQUIRED	TIME OF YEAR PROPAGATED	CULTURAL NOTES
Beaumontia	SEED	Sow in a temperature of 21–24°C. Grow on in the greenhouse border.	In spring	A flowering climber for a cool greenhouse. Can also be raised from cuttings.
Beloperone (*B. guttata = Shrimp Plant*)	TIP CUTTINGS	Root 5 cm cuttings in a propagator at 21°C.	March to July	Perennial houseplant up to 30cm in height. 16°C in good light.
Billbergia	DIVISION (offsets: suckers)	Remove large offsets from around base after flowering. Treat as cuttings rooting in a propagator at 24–29°C.	April	Evergreen houseplants (*B. nutans*) of the bromeliad family. Grow cool 7–10°C. The larger the offset when removed the better.
Boronia	SEED	Sow in a temperature of 18–21°C. Grow on in lime-free soil or compost.	March to May	Flowering aromatic shrubs for a cool greenhouse in border or 18 cm pots. 7°C.
Bougainvillea	HALF-RIPE CUTTINGS	Root 10 cm cuttings in sand/peat in a warm propagator.	August	Climbing deciduous plants for a cool greenhouse in pots or borders. 7–10°C.
Browallia	SEED	Sow in a temperature of 16–18°C. Will need a 13 cm pot for its final development.	February or July	Flowering perennials grown as greenhouse annuals good in the house when in flower. 7°C.
Brunfelsia	SEED	Sow in a temperature of 21–24°C.	In spring	Small evergreen flowering shrubs for the greenhouse. 10°C.
	HALF-RIPE CUTTINGS	Root in a propagator at 18–21°C.	June/July	
Caesalpinia (*Paradise Bird Flower*)	SEED (warm water soak 48 hours before sowing)	Sow in a temperature of 18–21°C. Can also be propagated by layering.	February to May	Flowering shrubs most suited to a greenhouse, 10°C in 25 cm pots. *C. gilliesii* and *C. japonica* can be grown outdoors in mild areas.
Caladium	DIVISION (of tuber)	Repot and divide large tubers before starting into growth in a propagator at 21–24°C.	March	Tuberous-rooted foliage plants up to 30 cm in height. Houseplant in summer, stored moist but not wet in winter. 16°C.
Calathea	DIVISION	Repot annually and divide, growing on singly in small pots.	March	Climbing foliage plants for a stove house in shade. 18°C. Smaller types make good bottle garden subjects.
Calceolaria (*Slipper Flower*)	SEED	Sow in a warm shady position in a greenhouse. Will need 10–13 cm pots for final development.	May to July	The tender types only need cool house treatment. 7°C. Houseplants when in flower.
Callistemon (*Bottle Brush Tree*)	SEED	Sow in a temperature of 16–18°C.	March to June	Evergreen flowering shrubs best placed in a cool house 7–10°C though *C. salignus* and *C. subulatus* are hardy in mild areas.
	HALF-RIPE CUTTINGS	Root 8 cm cuttings in a propagator at 16°C.	August	
Canna (*Indian Shot*)	SEED (chip and soak 24 hours in tepid water before sowing)	Sow in a temperature of 18–24°C. Use 18–20 cm pots for flowering under glass.	March to May	Herbaceous perennials for a cool greenhouse or outdoors during summer. Store tubers dry and frost-free in winter.
	DIVISION	Repot divisions in small pots.	March	
Capsicum (*Ornamental Pepper*)	SEED	Sow in a temperature of 16–18°C. Pot on into 9 cm for final potting.	February/March or May/June	*C. frutescens* is grown for its colourful fruits. Cool greenhouse or as a houseplant.
Carex (*Blue Grass*)	DIVISION	Divide when repotting at 3 year intervals.	March	Small foliage houseplant requiring a minimum of 10°C in winter.
Carica (*C. papaya = Common Papaw*)	SEED (soak 24 hours in tepid water before sowing)	Sow in a temperature of 21–24°C. Can also be raised from cuttings.	February onwards	Evergreen trees for a cool house mostly grown for its foliage though it will bear fruit. 7–10°C.

PLANT NAME	METHOD OF PROPAGATION	TREATMENT REQUIRED	TIME OF YEAR PROPAGATED	CULTURAL NOTES
Cassia *(C. alata = Candle Bush)*	SEED HALF-RIPE CUTTINGS	Sow in a temperature of 21–24°C. Root in a warm shady propagator.	March to May June	Small to medium-sized evergreen shrubs and perennials. Some are hardy in mild areas but most need a cool greenhouse.
Casuarina *(She-Oak:* *C. equisetifolia =* *Horsetail Tree)*	SEED HALF-RIPE CUTTINGS	Sow in a temperature of 16–18°C. Root in a shady propagator.	In spring July/August	Evergreen trees and shrubs, a number hardy in very mild areas but most needing frost-free conditions.
Chamaerops	SEED (Soak in warm water 48 hours before sowing)	Sow in a temperature of 18–24°C.	March to May	Evergreen palm up to 1.8 m in height. Outdoors in mild areas, as a pot plant in frost-free greenhouse elsewhere.
Chlorophytum *(C. elatum variegatum,* *C. comosum* *variegatum = Spider* *Plant)*	DIVISION (of roots) DIVISION (offsets: plantlets)	Divide and repot in 8 cm pots. Peg offsets down on pots of compost as roots begin to form.	In spring As they appear	Ideal foliage houseplants only requiring frost-free temperatures. Grow in 13 cm half pots.
Cineraria *(Senecio cruentus)*	SEED (surface sow)	Sow in a temperature of 16–18°C. Grow on in shady cool but frost-free conditions. 15 cm pot in final potting.	May to August	Pot plants flowering some 6 months from sowing. Good houseplant when in flower after which they are thrown away.
Cissus *(C. antarctica =* *Kangaroo Vine)*	HALF-RIPE CUTTINGS	Root 5 cm cuttings with a heel in a warm propagator.	During late spring/early summer	Evergreen climbing foliage plants ideal for the house. 10–13°C.
Citrus *(Orange etc; C. mitis is* *most commonly grown;* *varieties are budded or* *grafted)*	SEED (pips) HALF-RIPE CUTTINGS	Sow in slight heat in a shady position. Root 8 cm cuttings in a warm propagator.	As soon as ripe or in spring July/August	Reasonable foliage houseplants when small. Grow on in large tubs in a frost-free greenhouse.
Clerodendrum *(C. splendens =* *Bleeding Heart Vine)*	HALF-RIPE CUTTINGS SEED	Use ripe shoots and root in a warm propagator. Sow in a temperature of 18–24°C.	During summer March	*C. splendens* is a flowering climber for the greenhouse. 13–16°C. Grow in a 18 cm pot.
Clianthus *(C. puniceus = Parrot's* *Bill)*	SEED HALF-RIPE CUTTINGS	Sow in a temperature of 16–18°C. Root in sand/peat in a warm propagator.	March to June July/August	Flowering evergreen climbing plants for a cold greenhouse in full sun. Prick out or pot on in 13 cm pots.
Clitoria	SEED	Sow in a temperature of 21–24°C.	March to June	Evergreen climbers up to 4.5 m for a warm greenhouse. 10–13°C.
Coccoloba *(C. uvifera = Seaside* *Grape)*	SEED	Sow in a temperature of 18–21°C. Grow on in a sandy compost.	March/April	Foliage houseplants eventually forming a tree with edible fruit. Frost-free conditions.
Codiaeum *(Croton/South Sea* *Laurel)*	SOFTWOOD TIP CUTTINGS	Dip short cuttings in charcoal to stop 'bleeding' and root in a propagator at 21–24°C, spraying frequently.	At any time	Foliage plants for a constant temperature no lower than 16°C in full sun. Make difficult houseplants. Can also be air layered.
Coffea *(Coffee Tree)*	SEED HALF-RIPE CUTTINGS	Sow singly in pots in a temperature of 24°C. Root in a warm propagator.	February to May July/August	Shrubby plants grown mostly for their foliage. Houseplant in shady place. 7°C.
Coleus *(C. blumei = Flame* *Nettle)*	SEED SOFTWOOD CUTTINGS	Sow in a temperature of 16–18°C. Make cuttings of 5–8 cm and root in slight heat.	February to June August	Foliage plants. Grow on only the brightest seedlings. Houseplant. Overwinter at 13°C or grow as an annual.

PLANT NAME	METHOD OF PROPAGATION	TREATMENT REQUIRED	TIME OF YEAR PROPAGATED	CULTURAL NOTES
Collinia (C. elegans, syn. Chamaedorea elegans/Neanthe bella)	SEED (soak 48 hours in warm water before sowing)	Sow in a propagator at 27–29°C.	February to April	Small slow-growing palm as houseplant or in bottle gardens while small. Good light. 10°C.
Columnea	TIP CUTTINGS	Root 5 cm cuttings in a peat/sphagnum moss/sand compost in a propagator at 21°C.	March to May	Trailing evergreen flowering plants good in baskets. Houseplants in good light but out of full sun. 13°C.
Cordyline	STEM CUTTINGS	Root 2.5–5 cm sections of stem horizontally in a propagator at 24–27°C.	March/April	Evergreen trees and shrubs grown for their foliage. *C. terminalis* is a houseplant for slight shade. 7°C. Do not overpot. Offsets can also be grown on.
	SEED	Sow in a temperature of 24–27°C.	March/April	
Crossandra	SEED	Sow in a temperature of 18–21°C.	March to May	Small flowering evergreen shrubs. Young plants are to be preferred. Warm greenhouse. 13°C.
	SOFTWOOD CUTTINGS	Root in a slightly heated propagator in sand/peat compost.	March	
Cryptanthus (Earth Star)	DIVISION (offsets/suckers)	Remove larger offsets and establish in a propagator at 24–29°C.	April	Small bromeliads grown for their foliage. Ideal for bottle gardens. 7–10°C.
Cunonia	SEED	Sow in a temperature of 21–24°C.	March to May	*C. capensis* is a flowering shrub up to 3 m in height. Cool greenhouse 7–10°C.
	HALF-RIPE CUTTINGS	Root in a slightly heated propagator in sand/peat compost.	During summer	
Cuphea (C. ignea = Mexican Cigar Plant)	SEED	Sow in a temperature of 18°C.	February/March	Flowering greenhouse perennial 10°C. Pot into 13–15 cm pots for final potting.
	SOFTWOOD CUTTINGS	Root 5 cm cuttings in a warm propagator.	March/April	
Cycas (Cycads)	SEED (large; slow germination)	Sow singly in a sterile compost as for fern spores in a temperature of 21–24°C.	March to May	Houseplant or greenhouse foliage plants looking similar to tree-ferns. 7–10°C.
Cyperus (C. alterifolius = Umbrella Plant)	LEAF CUTTINGS	Cut mature leaf with 15 cm of stalk and root inverted in a jar of water. Pot into compost when rooted.	During spring and summer	Elegant foliage houseplants best stood in water permanently. Some species are hardy but tender species need 7–10°C in winter.
	SEED	Sow in a propagator at 18–21°C.	March	
Cyphomandra (C. betacea = Tree Tomato)	SEED	Sow in a temperature of 21–24°C.	March to May	Shrubby perennial up to 3 m in height. Bears edible fruits. 10°C.
Datura (Trumpet Flower. D. arborea = Angel's Trumpet)	SEED	Sow in a temperature of 18–21°C.	February to May	Shrubby types are medium-sized flowering shrubs for large pots or tubs in a greenhouse. 7–10°C. Some species are HHAs.
	STEM CUTTINGS	Root 15 cm cuttings in a propagator at 18°C.	April or September	
Desmodium (D. gyrans = Telegraph Plant)	SEED	Sow in a propagator at 18–21°C. Pot in a 13 cm for final potting.	March/April	A greenhouse annual grown for the peculiar gyrations of its leaves in high temperatures. Some species are hardy.
Dichorisandra (Seersucker Plant)	DIVISION	Divide when repotting and repot in a peat based compost.	March	Greenhouse flowering foliage plant for warm moist conditions. 16°C.
Dieffenbachia (Dumb Cane)	STEM OR TIP CUTTING	Root 8 cm tip or sections of the stem laid horizontally in a propagator at 24–27°C.	During spring or summer	Poisonous evergreen foliage plant for the warm greenhouse or as difficult houseplant in moist atmosphere. 13°C.

PLANT NAME	METHOD OF PROPAGATION	TREATMENT REQUIRED	TIME OF YEAR PROPAGATED	CULTURAL NOTES
Dionaea (D. muscipular = Venus' Fly Trap)	SEED	Sow on peat in a shady propagator. Keep moist and cover with glass until germinated.	March/April	Insectivorous bog plants for sun or shade and a cool but frost-free environment. Peat/live sphagnum compost with pots stood in water.
	DIVISION	Replant in 8 cm pots.	March	
Dizygotheca (D. elegantissima = False or Finger Aralia)	SEED	Sow in a temperature of 21–27°C.	March	Slow-growing foliage plant forming a shrub 60–90 cm in height. Difficult houseplant; best in warm greenhouse 16°C.
	SOFTWOOD CUTTINGS	Root side-shoots in a propagator at 18–21°C	July/August	
	ROOT CUTTINGS	Root in a propagator at 18–21°C.	March	
Dracaena (Dragon Plant)	SEED	Sow in a temperature of 24–29°C.	March to May	Evergreen foliage houseplants or summer bedding in mild areas. Sun or slight shade. 10–13°C.
	BUD OR TIP CUTTINGS	Root in a propagator at 16–21°C.	March/April	
Drosera (Sundew)	SEED	Sow on a peat/sphagnum compost in a propagator at 18°C.	March to May	Insectivorous perennial, some hardy but many needing a cool but frost-free greenhouse. Stand pots in a little water.
	DIVISION	Repot into a sedge peat/sphagnum moss compost.	In spring	
Duranta (Forget-me-not Tree)	SEED	Sow in a temperature of 21–24°C. Can also be raised from cuttings.	March to May	Evergreen flowering shrubs up to 3 m in height for a very warm greenhouse. 16°C.
Elettaria (Ginger Plant)	DIVISION	Divide and repot into JI potting No. 2.	At any time	Foliage houseplant up to 30 cm in height. Slight shade. 13°C.
Episcea (Trailing African Violet)	SOFTWOOD CUTTINGS	Root 5 cm cuttings in a propagator at 21–24°C.	In spring	Trailing flowering plants for a warm moist greenhouse 16°C.
Eranthemum (Lonely Flower)	TIP CUTTINGS	Root 5 cm cuttings in sand/peat in a propagator at 21–24°C.	April to June	Tall flowering foliage plants for a warm greenhouse 13–16°C.
Euphorbia (E. pulcherrima = Poinsettia)	TIP CUTTINGS	Root 5 cm cuttings in a propagator at 27–29°C, dipping cuttings in charcoal to seal them before insertion.	May/June	Shrubby plants with colourful bracts around flower. Greenhouse 13–16°C or houseplant while in 'flower'.
Exacum	SEED	Sow in a ternperature of 18–24°C.	March	Greenhouse flowering annual. Grow in warm, moist shady position.
Fatshedera (a hybrid between Fatsia japonica and Hedera helix)	TIP CUTTINGS	Root in a propagator at 24°C.	In spring	A foliage houseplant for cool conditions. Best kept frost-free in winter. Sun or shade.
Fatsia (F. japonica, syn. Aralia sieboldii = Fig Leaf Palm)	SEED	Sow in a temperature of 18–24°C.	March/April	Evergreen foliage shrub for cool shady conditions. Houseplant, cold house or outdoors in mild areas.
	HALF-RIPE CUTTINGS	Root in a warm propagator.	August	
Feijoa (F. sellowiana = Fruit Salad Bush)	SEED	Sow in a temperature of 21–24°C.	March to May	Evergreen shrub or small tree for frost-free greenhouse or warm wall in mild areas. Fruit and flowers are edible.
	HALF-RIPE CUTTINGS	Root in a warm propagator	July/August	
Felicia (Cape Aster; F. amelloides = Blue Marguerite)	SEED	Sow in a temperature of 16–18°C.	March/April	Shrubby perennials up to 45 cm in height for a cool greenhouse 7–10°C or outdoors in summer. Some species are HHAs.
	SOFTWOOD CUTTINGS	Root in a propagator at 16°C.	July/August	
Ficus (F. decora = Rubber Plant; F. benjamina = Weeping Fig; F. pumila = Climbing Fig)	SEED	Sow in a temperature of 18–21°C.	February/March	Foliage houseplants with a number of diverse species. The trailing/climbing F. pumila can be layered into small pots at any time. Keep fairly dry in winter. 10°C.
	AIR LAYERING (F. decora)	Cut a tongue and layer (see page 29).	May to July	
	HEEL CUTTINGS (F. benjamina)	Root in sand/peat in a propagator at 21–24°C.	During summer	

PLANT NAME	METHOD OF PROPAGATION	TREATMENT REQUIRED	TIME OF YEAR PROPAGATED	CULTURAL NOTES
Fittonia	HALF-RIPE TIP CUTTINGS	Root short cuttings in a propagator at 24°C.	March	Creeping foliage plants for warm, moist shady positions. Ideal for bottle gardens. 13°C.
Francoa (F. sonchifolia = Bridal Wreath)	SEED	Sow in slight heat.	March/April	Flowering perennials some hardy in mild areas but best in a cool greenhouse 7–10°C.
	DIVISION	Repot annually and divide. Need a 13–15 cm pot for final potting.	In spring	
Fuchsia (Lady's Eardrops)	SEED (surface sow)	Sow in a temperature of 21–24°C. and grow on in shady place under glass.	February to April	Many types and varieties for greenhouse as houseplants and some outdoors in mild areas. Only require frost-free conditions in winter.
	SOFTWOOD CUTTINGS	Root small cuttings in a warm propagator.	In spring	
Gardenia (G. jasminoides = Cape Jasmine)	SEED	Sow in a temperature of 21–24°C.	March to May	Evergreen flowering shrubs for a warm greenhouse. 13–16°C. Can be stood outdoors in a warm spot in summer.
	SOFTWOOD CUTTINGS	Root 8 cm cuttings in sand/peat in a warm propagator and establish in heat.	March	
Gerbera (G. jamesonii = Transvaal Daisy)	SEED	Sow shallowly in a temperature of 21°C	January to March	Flowering plants usually grown in the greenhouse border. Only requires frost-free conditions. Do not overwater.
	DIVISION (of crown)	Replant with the top of crown at soil level or a little above.	June	
Gossypium (G. herbaceum = Cotton Bush)	SEED	Sow in a temperature of 18–21°C. Pot into 13 cm for final potting.	March to May	Flowering perennial for a warm greenhouse. 10°C. 1–1.2 m in height. Fruit contains cotton.
Grevillea (G. robusta = Silk Bark Oak)	SEED	Sow in a temperature of 18–21°C.	February to June	Small shrubs to tall trees. Evergreen and flowering. Sun or shade. Lime-free soils or compost. Mostly 10°C although some are almost hardy.
	HALF-RIPE HEEL CUTTINGS	Root 8 cm cuttings in a warm propagator.	June/July	
Gynura	SOFTWOOD CUTTINGS	Root in a warm shady propagator and grow on in JI potting No. 1.	During spring and summer	Foliage houseplant for light position but out of sunshine. Root annually. 13°C.
Hardenbergia	SEED	Sow in a temperature of 21–24°C.	March to May	Evergreen flowering climber up to 2.4 m for frost-free greenhouse or a warm wall in mild areas.
	HALF-RIPE CUTTINGS	Root in a warm propagator.	August	
Hedera (Ivy)	SOFTWOOD OR HALF-RIPE CUTTINGS	Make cuttings 5–8 cm long and root in a warm propagator.	June/July	Hardy or slightly tender (H. canariensis) climbing or trailing evergreen shrubs with many varieties suitable as houseplants. Grow in slight shade.
	LAYERING	Peg down appropriate shoots indoors or out wherever plants are growing.	At any time	
Hedychium (Ginger Lily)	SEED	Sow in a temperature of 21–24°C.	February/March	Flowering herbaceous perennials for a warm greenhouse 10°C. or summer bedding. H. gardnerianum is the hardiest.
	DIVISION (of rhizome)	Divide when repotting and grow in JI potting compost No. 1.	In spring	
Helxine (H. soleirolii = Mind-your-own-business)	DIVISION	Replant in pots or under greenhouse staging where it makes a pleasant ground cover.	At any time or in April when repotting.	Small creeping houseplant for cool shady positions. Keep roots always moist but foliage dry. Is frost hardy if foliage dry.
Howea (syn. Kentia, Palm)	SEED (warm water soak 48 hours before sowing)	Sow in a temperature of 27–29°C.	February to April	Palm up to 1.2 m in height as houseplant out of direct sunlight. 13–16°C.
Hoya (H. carnosa = Wax Flower)	HALF-RIPE CUTTINGS	Root cuttings of previous year's growth in a warm propagator. Can also be layered.	March/April	Succulent flowering evergreen houseplants. H. bella is hanging, H. carnosa is climbing or trailing. 13°C.

PLANT NAME	METHOD OF PROPAGATION	TREATMENT REQUIRED	TIME OF YEAR PROPAGATED	CULTURAL NOTES
Humea (H. elegans = Incense Plant)	SEED	Sow in a temperature of 16°C. Grow on in 8 cm pots.	April/May	Tall biennial foliage/flowering greenhouse plant. 7–10°C.
Hypocyrta (H. glabra = Clog Plant)	HALF-RIPE CUTTINGS	Root in a warm propagator using 8 cm cuttings. Pot in Jl potting compost No. 1.	June to August	Low-growing foliage/flowering houseplants. 13°C. Water only when compost is dry.
Hypoestes (H. sanguinolenta = Polka-dot Plant)	SEED	Sow in a temperature of 18–21°C. 13 cm pot for final development.	March to May	Foliage houseplants for warm light positions. 13°C.
Impatiens (I. walleriana, syn. sultanii = Busy Lizzie)	SEED / SOFTWOOD CUTTINGS	Sow in a temperature of 21–24°C. / Root short cuttings in a propagator or in jar of water in a warm shady spot.	January/February / During summer	Very easy flowering houseplants or for summer bedding. 13°C. Do not overwater in winter, and grow in good light.
Indigofera	SEED (Soak in hot water till cool then sow) / HALF-RIPE CUTTINGS	Sow singly in small pots in a temperature of 21–24°C. / Root in a warm propagator.	March to May / July/August	Small, flowering, deciduous shrubs for frost-free greenhouse though some are hardy in mild areas. Good pot plants.
Ixora	HALF-RIPE CUTTINGS	Root 5 cm firm shoots in a propagator at 24°C.	May to August	Flowering evergreen shrubs or perennials for a warm greenhouse. 13°C.
Jacaranda	SEED / HALF-RIPE CUTTINGS	Sow in a temperature of 18–21°C. / Root in sand/peat in a warm propagator.	February to June / July/August	Large, flowering evergreen shrubs or trees mostly grown for their fern-like foliage. 7°C.
Jacobinia	SOFTWOOD CUTTINGS	Root in peat/sand in a warm propagator.	April to June	Small to medium-sized flowering plants for a warm greenhouse. 13°C.
Lagerstroemia (L. indica = Crape Myrtle)	SEED / SOFTWOOD CUTTINGS	Sow in a temperature of 21–24°C. / Root in a warm propagator.	March / April to June	Flowering deciduous trees and shrubs best in greenhouse borders, 7°C, or in large tubs stood outdoors during summer.
Lagunaria	SEED	Sow in a temperature of 21–24°C.	March to May	Evergreen flowering tree up to 6 m in height. Frost-free greenhouse or outdoors in mild areas.
Lantana	SEED / SOFTWOOD CUTTINGS	Sow in a propagator at 21–24°C. / Root in a warm propagator. Also half-ripe cuttings in August or September.	February/March / In spring	Flowering evergreen shrubs for a greenhouse 7–10°C, or as summer bedding. Can be pruned to keep in shape.
Lapageria (L. rosea = Chilean Bell Flower)	SEED / LAYERING	Sow in a temperature of 21–24°C. / Cut a tongue and peg down into small pots of compost.	March to May / Spring through summer	Flowering evergreen climber best in a lime-free compost in a greenhouse border 7–10°C. Slight shade.
Leea (L. coccinea = West Indian Holly)	SEED	Sow in a moist atmosphere at a temperature of 24–27°C.	March to May	L. coccinea is a houseplant grown for foliage, flowers and fruit. 13–16°C.
Licuala	SEED (soak 48 hours in warm water before sowing)	Sow in a temperature of 24–27°C.	March to May	Medium-sized palms for a warm greenhouse. 13–16°C.
Lomatia	SEED / HALF-RIPE CUTTINGS	Sow in a temperature of 21–24°C. / Root in sand/peat in a slightly heated propagator.	February to May / August/September	Evergreen shrubs grown as foliage plants in cool greenhouses 7–10°C. L. ferruginea is hardy in mild areas.

PLANT NAME	METHOD OF PROPAGATION	TREATMENT REQUIRED	TIME OF YEAR PROPAGATED	CULTURAL NOTES
Lotus (*L. berthelotti = Coral Gem*; *L. corniculatus = Bird's Foot Trefoil*; *HP*)	SEED DIVISION (hardy types)	Sow in a temperature of 18°C. Replant in lime-rich soils in the rock garden.	April In spring	Flowering perennials, *L. berthelotti* being a cool greenhouse plant whereas others are hardy. 7°C.
Luculia	SEED	Sow in a temperature of 21–24°C. Can also be raised from summer cuttings.	March to May	Winter-flowering deciduous shrubs best in pots or borders in a cool greenhouse. 7°C.
Luffa (*Loofah*)	SEED	Sow singly 13 mm deep in peat pots in a propagator at 18–21°C. Grow on in cooler but very humid atmosphere.	February/March	Annual greenhouse climbing plants similar to and cultivated as for the cucumber. Fibrous interior of ripe fruits produce the 'loofahs'.
Mandevilla (*M. suaveolens = Chilean Jasmine*)	SEED HALF-RIPE CUTTINGS	Sow in a temperature of 21–24°C. Root 8 cm cuttings of side-shoots in a slightly heated propagator.	March to May July/August	Flowering deciduous climbers for the borders of frost-free greenhouse or conservatory.
Maranta (*Prayer Plant*; *M. arundinacea = Arrowroot*)	DIVISION	Replant in small pots to grow on but do not overpot.	March	Mostly small foliage plants for shady moist atmospheres. Good in bottle gardens. 13°C.
Medinilla	HALF-RIPE HEEL CUTTINGS	Root in a propagator at around 21°C. Pot into 15 cm pots for final development.	November	Greenhouse shrubs. 10–13°C. *M. magnifica* is an evergreen flowering shrub up to 90 cm in height.
Melianthus (*M. major = Cape Honey Flower*)	SEED HALF-RIPE CUTTINGS	Sow in a temperature of 21–24°C. Root in a warm propagator.	February to April June to August	Evergreen shrubs up to 1.5 m in height. Mostly grown for their foliage. 7–10°C except *M. major* which is hardy in mild areas.
Microsperma	SEED	Sow in a temperature of 16–18°C.	February to April	Small flowering annual for the greenhouse or summer bedding in warm areas.
Mimosa (*Sensitive Plant = M. sensitiva*, *M. padica*)	SEED	Sow in a temperature of 18°C and grow on in 8–10 cm pots.	February to April	Leaves sensitive to touch. Perennial but best as a warm greenhouse annual.
Monstera (*Shingle Plant. M. deliciosa = Swiss Cheese Plant*)	TIP CUTTINGS SEED	Root in a propagator at 24°C. Sow 13 mm deep in a peat-based compost in a moist propagator at 24–29°C.	During spring and summer June or as soon as available	Climbing foliage houseplant with aerial roots. 16°C. Keep out of direct sunlight in summer.
Musa (*Banana*)	SEED (soak 24 hours in warm water before sowing) DIVISION (suckers)	Sow singly in small pots in a temperature of 27°C. Remove suckers, pot, and establish in heat. 24°C.	February to May At any time	Herbaceous foliage plants for slight heat or sub-tropical bedding. 25 cm pot indoors and 7–10°C in winter. *M. basjoo* is the hardiest.
Neoregelia	SEED DIVISION (offsets)	Sow in a temperature of 21–27°C. Remove when large enough, pot, and grow on in 21°C.	March to May As formed	Bromeliads grown as foliage houseplants. 16°C. Keep central vase filled with rain water.
Nepenthes (*Pitcher Plant*)	SEED (surface sow) RIPE STEM CUTTINGS	Sow on peat/sphagnum moss in a propagator at 24–29°C. Root 8–15 cm cuttings in peat/sphagnum moss in a propagator at 27°C.	In spring May to August	Insectivorous plants for a very warm greenhouse. 18°C. Shade from direct sun and pot in a compost of fibrous peat, sphagnum moss, coarse sand and a little charcoal.

PLANT NAME	METHOD OF PROPAGATION	TREATMENT REQUIRED	TIME OF YEAR PROPAGATED	CULTURAL NOTES
Nerium *(N. oleander = Oleander)*	SEED TIP CUTTINGS	Sow in a temperature of 21–24°. Root 5 cm cuttings in permanently moist sand in a warm propagator.	March to May July/August	Evergreen flowering shrubs for frost-free greenhouses in full sun. Poisonous in all parts to man and beast.
Nertera *(N. granadensis = Bead Plant)*	SEED (soak 12 hours in warm water before sowing) DIVISION	Sow in peat/sand in a propagator at 18–21°C. Repot small divisions and establish in heat.	March to May In spring	*N. granadensis* is a berry-bearing trailing plant for a frost-free greenhouse or rock garden with overhead protection in mild areas.
Nidularium *(Bird's Nest Bromeliad)*	DIVISION (offsets)	Remove large offsets and pot singly in sand/peat. Establish in a propagator at 27°C.	April to July as available	Bromeliad grown as an evergreen foliage houseplant. Shady positions. Keep central vase filled with rain water. 13°C.
Nierembergia	SEED HALF-RIPE CUTTINGS	Sow in a temperature of 16°C. Root in sand/peat in a warm propagator.	March to May July/August	Greenhouse types need frost-free conditions. They are low-growing flowering perennials.
Orchids *(Greenhouse types)*				A vast range of genera, species and varieties whose diversity of cultural requirements only leave room for a simplified outline of a few representative types. Many are not difficult providing that their special needs are met. Orchids are usually grown in special greenhouses often restricted to a single genus. *Cymbidiums* are the most widely grown and some can be grown in a mixed cool house 7°C or as houseplants with care. For beginners the assistance of an experienced grower at first never goes amiss. Compost for many types: equal parts coarse fibrous peat/perlite and chopped sphagnum moss with a very small amount of slow release fertiliser. Some orchids are also raised from cuttings and commercially by meristem culture. The necessary apparatus and instructions for artificial seed raising are available in kit form from specialists.
Cymbidiums *(cool or intermediate house 7°C)*	DIVISION (of pseudo-bulbs)	Leave 3 inactive back bulbs behind new growth when repotting, cutting away any spare and potting these singly in small pots. Grow on in a warm, moist propagator.	February to April	
Dendrobiums *(deciduous types: intermediate house 10°C)*	DIVISION (offsets: plantlets)	Carefully remove plantlets from leaf-nodes of deciduous types when roots are 5 cm or so long. Pot singly in 2 parts fibre/1 sphagnum moss.	As they become available	
Paphiopedilum *(Cypripedium are the hardy types: intermediate house 13°C)*	DIVISION (of plant)	Repot annually and divide plants with 4 to 6 growths if desired. Repot in 2 parts fibre/1 sphagnum moss/1 fibrous loam with some charcoal and drainage material. Spray over but do not water for a few weeks.	After flowering	
All species	SEED (natural germination needs presence of symbiotic fungus)	Usually raised artificially in an agar jelly in sterile conditions although some success can be achieved by sowing on the prepared (trimmed) compost of established plants of a related genus. Any seedlings appearing are potted on after 1 year.	At any time	
Pandanus *(Screw Pine)*	SEED DIVISION (offsets: suckers)	Sow in a temperature of 24–27°C. Repot shallowly in lime-free compost and establish in heat.	March to May In spring	Palm-like foliage plants best in a warm, moist greenhouse or as a houseplant when young. 13°C.
Passiflora *(Passion Flower)*	SEED HEEL CUTTINGS	Sow in a propagator at 18–21°C. Root 15 cm cuttings in a warm propagator.	March/April May to July	Climbing flowering perennials. *P. caerulea* is hardy in mild areas or for cold greenhouses. Other species need more heat.
Pavonia	SEED SOFTWOOD OR HALF-RIPE CUTTINGS	Sow in a temperature of 18–21°C. Root in sand in a propagator at 18–24°C	February to April May to August	Shrubby evergreen flowering perennials for a greenhouse 7–10°C in winter.

PLANT NAME	METHOD OF PROPAGATION	TREATMENT REQUIRED	TIME OF YEAR PROPAGATED	CULTURAL NOTES
Pellionia	DIVISION	Replant in small pots.	April	*P. daveauana* is a trailing foliage houseplant for warm, moist conditions. 13–16°C.
	SOFTWOOD CUTTINGS	Root in a warm moist propagator.	June to August	
Pentas (Star Cluster)	SEEDS	Sow in a temperature of 21–24°C.	March/April	Small shrubby flowering plants for a warm greenhouse. Up to 60 cm in height. 13–16°C.
	SOFTWOOD CUTTINGS	Root in peat/sand in a propagator at 16–18°C.	April/May	
Peperomia (Pepper Elder)	LEAF CUTTINGS (most types)	Root with leaf and stalk in a propagator at 18°C.	During spring and summer	Mostly low-growing good foliage houseplants 7–10°C. Water sparingly with tepid water.
	SOFTWOOD CUTTINGS	Root in a propagator at 18°C.	April to August	
Persea (P. gratissima = Avocado Pear)	SEED	Plant singly in a 13 cm pot pointed end up in a warm shady place. Can also be grown from cuttings.	During spring and summer	Easily raised from shop bought fruit stones. Makes reasonable foliage houseplant. 10°C.
Petrea (Purple Wreath)	SEED	Sow in a temperature of 24°C.	March to May	Large flowering climbing shrubs for borders or 25 cm pots in a warm 10°C greenhouse.
	HALF-RIPE CUTTINGS	Root 5 cm cuttings in peat/sand in a warm propagator.	May to August	
Philodendron	SEED	Sow in a temperature of 24°C.	March to May	Evergreen foliage and sometimes climbing houseplants. 13°C.
	SOFTWOOD CUTTINGS	Root short cuttings in a warm propagator.	At any time	
Phoenix (P. roebelenii = Dwarf Date Palm)	SEED (soak 48 hours in warm water before sowing)	Sow in moist peat at 27°C and pot into 13 cm pots as soon as growing.	In spring	Palms needing only frost protection in winter and good as houseplants.
Pilea (P. microphylla = Artillery Plant. P. cadierei = Aluminium Plant)	SEED	Sow in a shady propagator at 21–24°C.	March to May	Small to medium-sized foliage houseplants. Use a peaty compost and grow in slight shade. 10–13°C.
	SOFTWOOD CUTTINGS	Root in slight heat in a peat-based compost.	During spring and summer	
Plectranthus	SEED	Sow in a temperature of 24°C.	March to May	*P. oertendahlii* is a trailing foliage houseplant and can be divided in April. Other species are greenhouse plants. 7–10°C.
	HALF-RIPE CUTTINGS	Root in a warm propagator.	July/August	
Plumbago (P. capensis = Cape Leadwort)	SEED	Sow in a temperature 18–21°C.	March to May	Flowering, climbing/trailing plants for a warm greenhouse. 7°C for *P. capensis* but some species require more.
	HEEL CUTTINGS (P. capensis)	Root short cuttings in a warm propagator.	In spring	
	ROOT CUTTINGS	Root 2.5 cm sections in peat/sand in a warm propagator.	December to February	
Plumeria (P. rubra = Frangipani)	SEED	Sow in a temperature of 21–24°C.	March to May	Fragrant flowering deciduous shrub or small tree for a warm greenhouse 13–16°C.
	SOFTWOOD CUTTINGS	Root in a sandy compost in a warm propagator.	April	
Podalyria (P. sericea = Silver Sweetpea Bush)	SEED	Sow in a temperature of 21–24°C. Pot on in a well-drained compost.	March to May	Low-growing evergreen shrubs best raised from seed. Grown under glass for foliage and flowers. 13°C.
Primula	SEED (surface sow)	Sow in a temperature of 16–18°C. Grow on in cooler conditions in slight shade. Grow on in JI potting compost No. 1.	February to June (P. kewensis) May to August (P. malacoides) February to June (P. obconica) May to July (P. sinensis)	Greenhouse annuals and/or perennial flowering plants. 13 cm pot for final potting. They can be stood out in frames for the summer and housed later. 7°C.

PLANT NAME	METHOD OF PROPAGATION	TREATMENT REQUIRED	TIME OF YEAR PROPAGATED	CULTURAL NOTES
Protea	SEED	Sow in a temperature of 21–24°C.	March/April	Small to medium-sized evergreen flowering shrubs for a cool greenhouse 7–10°C. Grow in pots or tubs and stand out in summer.
	HALF-RIPE CUTTINGS	Root in peat/sand in a warm, moist propagator.	July/August	
Psidium *(P. cattleianum = Strawberry Guava)*	SEED	Sow in a temperature 18–21°C.	March to May	Evergreen shrubs for a frost-free greenhouse. Grown for foliage and edible fruits. Grows well in pots.
	HALF-RIPE CUTTINGS	Root in a warm propagator in sand.	July/August	
Rehmannia	SEED	Sow in a temperature of 16°C.	June/July	Flowering pot plant best treated as biennial in a 13 cm pot. 7–10°C.
	BASAL CUTTINGS	Root in a cold frame.	After flowering	
Rhoeo *(Boat Lily)*	SEED (30 minute soak before sowing)	Sow in a temperature of 18–21°C.	March to May	Foliage houseplant 30 cm in height. Water well and shade in summer. 10°C.
	DIVISION (offsets)	Treat offsets as cuttings and root in slight heat.	June/July	
Rhoicissus *(R. rhomboidea = Grape Ivy)*	HALF-RIPE CUTTINGS	Root short cuttings in a warm propagator.	July/August	Evergreen foliage houseplants ideal for shady positions. 10–13°C.
Ruellia	SEED	Sow in a temperature of 21°C.	March to May	Flowering perennials and sub-shrubs many making good pot plants. 13°C but *R. ciliosa* is hardy in mild areas.
	STEM CUTTINGS	Root in a propagator at 24°C. Pot on in JI potting compost No. 2.	June to August	
Saintpaulia *(African Violets)*	SEED (Fl hybrids: surface sow)	Sow thinly in a temperature of 18–21°C. Grow on in warmth.	February to April	Houseplants flowering in 9 cm pots. Keep foliage dry only watering from below. 10°C. Grow in JI No. 1 with added peat.
	LEAF CUTTINGS	Use mature leaves with stalk and root in a propagator at 16°C.	During summer	
Sansevieria *(S. trifasciata laurentii = Mother-in-Law's Tongue)*	DIVISION (offsets)	Remove offsets when 20 cm high, pot 2–3 to a pot and grow on in warmth.	As they form	Foliage houseplant requiring a soil-based compost. Do not overwater. 7–10°C. *S. t. laurentii* leaf cuttings will revert to mottled green form.
	LEAF CUTTINGS	Cut leaves into 5 cm sections and treat as cuttings in a warm propagator.	May to August	
Sarracenia *(Indian Cup/Pitcher Plant)*	SEED (may benefit from cooling before sowing)	Sow on a constantly moist compost of peat/sphagnum moss at 13–16°C.	March/April	Insectivorous plants for cool shady moist positions 4°C, though some will grow in bog gardens outdoors in mild areas if protected overhead in winter.
	DIVISION	Replant in a compost 2 part fibrous peat/1 sphagnum moss.	In spring	
Schefflera *(S. actinophylla syn. Brassaia actinophylla = Umbrella Tree)*	SEED	Sow in a temperature of 18–21°C.	March to May	Slow-growing foliage houseplant. 13°C. Can also be raised from cuttings in June or July.
	AIR LAYERING	Layer as described in chapter 4.	May to August	
Scindapsus *(Giant Leaf/Devil's Ivy)*	TIP CUTTINGS	Root short cuttings in a warm propagator.	June to August	Evergreen foliage plants mostly climbing and best grown on a mossed stake. Houseplant. 10°C.
	BUD CUTTINGS	Root as above.	June to August	
Selaginella *(Creeping moss)*	LAYERING	Layer *in situ* or lay on moist compost and cover with glass using 8 cm pieces as cuttings. Can also be raised from spores.	During spring and summer	Ferny moss-like trailers for shady moist positions in warm greenhouses 13–16°C though some need more and some less. Equal parts fibrous peat and sphagnum moss compost.
Senecio *(S. macroglossus variegatus = German Ivy)*	HALF-RIPE CUTTINGS	Root in a warm propagator. Grow on in JI No. 2 and do not overwater.	July/August	*S. mikanioides* and *S. microglossus* are climbing houseplants similar to ivy. 7°C.

PLANT NAME	METHOD OF PROPAGATION	TREATMENT REQUIRED	TIME OF YEAR PROPAGATED	CULTURAL NOTES
Sesbania	SEED	Sow in a temperature of 18–21°C.	March to May	Mostly small to medium-sized flowering shrubs for warm, moist greenhouses. 16°C.
	HALF-RIPE CUTTINGS	Root short cuttings in a propagator at 24–27°C	During summer	
Setcreasea (Purple Heart)	DIVISION	Divide when repotting.	April	*S. purpurea* is a trailing foliage houseplant for good light. 10°C.
	SOFTWOOD CUTTINGS	Root in a warm shady spot in house or greenhouse.	Spring through summer	
Solanum (*S. capsicastrum* = Winter Cherry)	SEED	Sow in a temperature of 16–18°C. Grown on outdoors in sun during summer.	February/March	Grown for its berries, which are slightly poisonous. Houseplant. 7°C.
Sparmannia (*S. africana* = African Hemp)	SEED	Sow in a temperature of 18–21°C.	March to May	Shrubby evergreen flowering plant for house or greenhouse. 7°C. Can be stood outside during summer.
	SOFTWOOD CUTTINGS	Root in a warm propagator and pot singly as soon as rooted.	April to June	
Spathiphyllum (Peace Lily)	SEED	Sow in a temperature of 24–27°C.	March to May	Flowering evergreen houseplant up to 30 cm in height. 13°C. Humid conditions.
	DIVISION	Repot in JI potting compost No. 2.	April	
Stenocarpus (Queensland Fire Wheel Tree)	SEED	Sow in a temperature of 16–18°C.	March to August	Evergreen foliage tree which flowers while young. Frost-free conditions or houseplant while small.
Stephanotis (Madagascar Jasmine)	SEED	Sow in a temperature of 24–27°C.	January to April	Fragrant flowering evergreen twining plants, houseplants in good light and 13–16°C in winter.
	HALF-RIPE CUTTINGS	Root short cuttings in sand/peat in a warm propagator.	In spring	
Strelitzia (*S. reginae* = Bird of Paradise Flower)	SEED (soak 30 minutes before sowing)	Sow in a temperature of 21–24°C. Transplant as soon as germination occurs. Grow on in JI potting compost No. 2.	February/March	Greenhouse perennial for full sun in pots or borders. 10°C. Older plants can be divided but seeds give better results.
Streptocarpus (Cape Primrose)	SEED (surface sow)	Sow thinly in a temperature of 18–21°C. Grow on in warmth.	January/February	Greenhouse flowering plants or as houseplants. 13°C or slightly less when not in flower. Grow in JI potting compost No. 2.
	LEAF CUTTINGS	Cut veins of a mature leaf and peg down on compost surface and place in a warm propagator.	During summer	
Syagrus (*S. weddeliana*, syn. *Cocos weddelianus*)	SEED (soak 48 hours in warm water before sowing)	Sow singly in small pots in a temperature of 24°C.	March/April	Elegant houseplant palm good in bottle gardens while small. 16–18°C.
Syngonium (Goosefoot Plant: syn. *Nephthytis*)	TIP OR STEM CUTTINGS	Root stem sections containing 2 leaves in a warm propagator.	June to August	Foliage houseplant trailing or climbing up mossed stake. In shade. 13–16°C.
Tacca (Bat Plant)	SEED (soak 72 hours in warm water before sowing)	Sow in temperature of 27–29°C.	February to April	Warm greenhouse perennials grown for foliage and peculiar flowers. Do not overwater in winter. 16°C.
	DIVISION (of rhizome)	Divide when repotting into JI potting compost No. 1.	In spring	
Tephrosia	SEED	Sow in a temperature of 21–24°C. Grow on in a peat based compost.	March to May	*T. grandiflora* is a small greenhouse shrub with red flowers. 10–13°C.
Tetrastigma (*T. voinierianum* = Chestnut Vine)	TIP OR BUD CUTTINGS	Root in a warm propagator.	June to August	Vigorous climbing houseplant if there is space! 10–13°C.
Tibouchina (Glory Bush)	SEED	Sow in a temperature of 21–24°C.	March to May	Evergreen greenhouse shrubs flowering while still small. Grow in full sun. 10°C.
	HALF-RIPE CUTTINGS	Root in a warm propagator.	July/August	

PLANT NAME	METHOD OF PROPAGATION	TREATMENT REQUIRED	TIME OF YEAR PROPAGATED	CULTURAL NOTES
Tillandsia (Air Plants)	DIVISION (offsets: suckers)	Separate and pot into small pots of sand/peat. Grow on in a moist propagator at 21–24°C. Can also be raised from seed.	June	Foliage houseplants which also flower. Epiphytes needing a compost of fibrous peat and sphagnum moss. Grow on bark or in baskets. 13°C.
Tolmiea (T. menziesii = Pick-a-back Plant)	DIVISION (offsets: plantlets)	Root small plantlets indoors or in a cold frame. Grow on in slight shade.	During summer	Hardy perennial used as a houseplant. Produces young plants at leaf joints.
Torenia (T. fournieri = Wishbone Flower)	SEED	Sow in a temperature of 18–21°C. Grow on in JI potting compost No. 1.	March/April	Short, flowering annuals for a fairly warm greenhouse. Plant 3–4 in a 13 cm pot.
Tradescantia and Zebrina (T. fluminesis = Wandering Jew; Zebrina = Zebra Plant)	SOFTWOOD CUTTINGS	Root in slight heat in moist compost or in jar of water.	Spring through summer	*Tradescantia* and *Zebrina* are closely related trailing foliage houseplants. 7°C.
Tweedia (T. caerulea, syn. Oxypetalum caeruleum)	SEED HALF-RIPE CUTTINGS	Sow in a temperature of 21–24°C. Root in a propagator at 21–24°C.	March to May June to August	Twining flowering climber best in the border of a warm greenhouse in full sun. 13–16°C.
Vinca (V. rosea = Madagascar Periwinkle; syn. Catharanthus roseus)	SEED	Sow in a temperature of 18–21°C. Pot finally into a 13 cm pot.	March to May	Greenhouse perennial usually grown as an annual. 13°C if overwintered.
Vriesia	DIVISION (offsets)	Remove large offsets and pot singly in small pots of peat/sand in a propagator at 24–29°C.	April to August as available	Bromeliad foliage/flowering houseplant. Keep central vase filled with rainwater. 13°C.
Zingiber (Ginger Plant)	DIVISION (of tuber)	Repot singly in JI potting compost No. 1.	February when repotting	Perennial flowering plants, tuberous roots providing ginger. Grow in shady greenhouses 13–16°C.

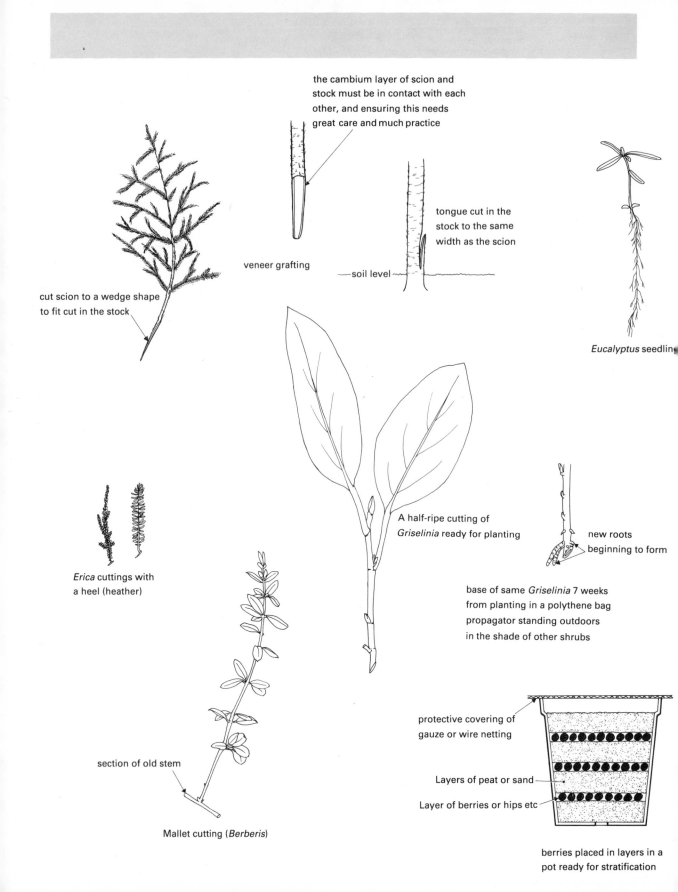

cut scion to a wedge shape
to fit cut in the stock

the cambium layer of scion and
stock must be in contact with each
other, and ensuring this needs
great care and much practice

veneer grafting

tongue cut in the
stock to the same
width as the scion

soil level

Eucalyptus seedling

Erica cuttings with
a heel (heather)

A half-ripe cutting of
Griselinia ready for planting

new roots
beginning to form

base of same *Griselinia* 7 weeks
from planting in a polythene bag
propagator standing outdoors
in the shade of other shrubs

section of old stem

Mallet cutting (*Berberis*)

protective covering of
gauze or wire netting

Layers of peat or sand

Layer of berries or hips etc

berries placed in layers in a
pot ready for stratification

18 Trees and Shrubs

The propagation of trees and shrubs is to me one of the most rewarding of gardening pursuits, for often such plants raised become 'friends' for life, and to watch them grow in stature from year to year is somehow deeply satisfying. From the enthusiastic propagator's point of view, too, there is something for him or her the whole year: collecting seed (often large) or other propagating material, sowing, rooting, transplanting, pruning—there is never a dull moment. It is surprising how many shrubs and trees can be accommodated in even a small garden, especially if use is also made of walls etc for the climbers.

As experience grows, more difficult subjects can be tackled, or perhaps a little grafting. Many conifers and other evergreens are grafted under glass in March and can provide a welcome task on a cold wet day. For many conifers a type of veneer graft is useful, this being a graft where the top remains on the potted two-year-old root-stock until a union is made. In the case of conifers, the scion is a piece of the previous year's growth, usually between 5 and 10 cm long, depending on the speed of growth of the donor. The slower the growth the shorter the scion.

The rootstock is prepared with a downward vertical cut just below the bark about 2.5 cm long and as near the compost surface as possible so as to produce a loose flap of bark with the wood just visible below. Cut the scion then with an equal-sided wedge cut at the base exactly matching the cut in the rootstock then push the wedge down behind the flap making sure that there is cambial contact on at least one side.

With the scion in place bind the wound with raffia and place the pot with rootstock and attached scion into a propagator with bottom heat until the union is formed. This takes some 6 weeks or so and after that the raffia can be removed and the rootstock beheaded before hardening the new plant off to outdoor conditions.

Many conifers, along with other trees and shrubs, will, however, root as cuttings, even the dwarf types which are often only grafted commercially to speed on their growth. The majority of evergreen cuttings will benefit from mist propagation, but the use of thin polythene laid just over the cuttings stood on a heated bench (18 °C) is a satisfactory alternative. Give the cuttings a drench of multi-purpose fungicide before covering and shade them according to the time of the year and weather conditions, i.e. heavily on a spring/summer day and lightly on cloudy days or when sunny earlier or later in the year.

I prefer to grow all summer/autumn rooted cuttings on in cold frames for their first winter before standing out or planting in nursery beds and in most cases root in a sand/peat compost. Resinous subjects such as *Cupressus* may benefit from a dip in acetone or boiling water before dipping in rooting powder to break down the resin at their base.

Vegetative propagation by cuttings, layering or grafting is necessary for most varietal forms, but the majority of the species can be raised from seed. Most hardy kinds are sown in drills in a seed bed outdoors, though I prefer to sow small amounts in pans in a cold frame for safety. Many will benefit from stratification (*see* pages 13 and 51). For stratifying before sowing use sand for dry seed and sand/peat for berries and other soft fruits.

Some tree and shrub seeds require a period of warmth followed by a period of cold before dormancy is broken, which accounts for many not germinating for a full year or more. This can be left to natural forces or can be recreated artificially, stratifying such seeds at 20–24 °C for 4–5 months followed by 3 at 4 °C.

The majority of seedlings are best transplanted to outdoor nursery beds after their initial pricking out into small pots, hardening off greenhouse-raised plants beforehand. Here they can remain for a few years before being placed in their final positions in autumn or winter. They can also be potted on, as are the liners at the garden centre, and I do this with small numbers—but one has to remember to water them!

In the following lists the subjects are hardy and will grow in ordinary soils in sunny positions unless stated.

PLANT NAME	METHOD OF PROPAGATION	TREATMENT REQUIRED	TIME OF YEAR PROPAGATED	CULTURAL NOTES
Abelia	HALF-RIPE CUTTINGS	Root in a warm propagator or under mist. Can also be layered in spring.	July/August	Small deciduous and evergreen shrubs best grown in milder areas especially A. floribunda which needs a warm wall or frost-free greenhouse.
Abeliophyllum	HALF-RIPE CUTTINGS	Root in a propagator at 18°C. Can also be layered in spring or autumn.	July/August	Small, flowering deciduous shrub hardy in a warm sheltered spot.
Abies (Silver Firs)	SEED	Store seeds at 4°C for 3 months then sow in drills outdoors.	March	Tall evergreen coniferous trees. Forms are veneer grafted onto species rootstock under glass in March.
Acer (Maple) A. japonicum and A. palmatum	SEED (stratify dry seed)	Sow in a seedbed outdoors.	Autumn as soon ripe or in spring	Small to medium-sized deciduous trees and shrubs grown for their foliage effects. Varieties are often grafted onto their respective species. Plant in sun or slight shade.
	SOFTWOOD CUTTINGS	Root in a propagator at 18°C.	April to June	
	BUDDING	Bud ordinary types (Maples) onto species. See Chapters 6 and 13.	June to August	
Actinidia (A. chinensis = Chinese Gooseberry)	SEED	Sow in pots and stratify outdoors. Move into slight heat in March.	January/February	Vigorous deciduous flowering climbers for trellises etc. The fruit of A. chinensis is edible.
	HALF-RIPE CUTTINGS	Root in a warm propagator.	July/August	
Aesculus (A. hippocastanum = Horsechestnut)	SEED	Sow 8 cm deep in a seedbed outdoors.	Autumn as soon as ripe	Small shrubs to large trees. Deciduous. Grown for their flowers and foliage. Can also be whip and tongue grafted outdoors in spring.
	BUDDING	Bud varieties onto seedling species using buds from the base of shoots.	July/August	
Ailanthus (A. altissima = Tree of Heaven)	SEED	Sow in a prepared bed outdoors.	March/April	Medium-sized trees mostly grown for their foliage, especially in industrial areas in the south.
	ROOT CUTTINGS	Root short sections in a cold frame or propagator.	December/January	
Akebia	SEED	Sow in slight heat.	September	Semi-evergreen climbing shrubs for sun or slight shade grown for their flowers and ornamental fruits.
	HALF-RIPE CUTTINGS	Root in a propagator with slight heat.	August	
	LAYERING	Layer appropriate shoots into pots or soil.	Late spring and summer	
Alnus (Alder)	SEED	Sow in pots and place out to stratify. Move into slight heat in March.	January/February	Medium-sized deciduous catkin-bearing trees and shrubs for sun or slight shade in moist soils. Variegated and other forms are grafted onto seedling species.
	LAYERING	Peg suitable branches down to root.	Through summer	
Amelanchier (A. canadensis = Snowy Mespilus/June Berry)	SEED	Sow in beds outdoors.	September/October	Small to medium-sized deciduous flowering trees and shrubs with good autumn colour. Grow in moist soils. Can also be layered.
	DIVISION (offsets: suckers)	Separate carefully and grow on in a nursery bed.	November to March	
Amorpha (A. canescens = Lead Plant)	DIVISION	Replant in sunny positions.	November to March	Small deciduous flowering shrubs and sub-shrubs. Can also be raised from seed when available.
	SOFTWOOD CUTTINGS	Root in a propagator in slight warmth.	May/June	
Aralia (A. elata = Angelica Tree)	ROOT CUTTINGS	Root and grow on in pots in a cold frame.	December/January	Medium-sized deciduous trees and herbaceous perennials, trees sometimes forming as shrubs. Grown for their foliage which may be damaged by hard frosts.
	SEED	Sow in a temperature of 16–18°C. May require warm/cool stratification.	March/April	

PLANT NAME	METHOD OF PROPAGATION	TREATMENT REQUIRED	TIME OF YEAR PROPAGATED	CULTURAL NOTES
Araucaria (A. araucana = Monkey Puzzle)	SEED	Sow in a temperature of 16–18°C.	March/April	Evergreen coniferous trees. A. araucana is hardy. Plant in a deep, rich, moist soil.
Arbutus (A. unedo = Strawberry Tree; bears edible fruits)	SEED	Sow 2.5 cm deep in pots in slight heat.	March/April	Small to large evergreen trees for sun or shade. Suitable for coastal areas especially.
	HALF-RIPE CUTTINGS	Root in a propagator at 18°C.	August/September	
Arundinaria (Bamboo)	DIVISION	Replant according to species in rich moist soils.	April/May	Shrubby grasses up to 4.5 m making good windbreaks or screens. A. japonica is one of the hardiest.
Atriplex (A. halimus = Tree Purslane)	HALF-RIPE CUTTINGS	Root in a cold frame or in a sheltered spot outdoors.	September/October	Evergreen or semi-evergreens with silver-grey foliage. Shrubby types make good hedges in coastal areas.
Aucuba (A. japonica = Spotted Laurel)	SEED	Sow in pots outdoors to stratify then move to slight heat March.	January/February	Medium-sized shrubs for sun or shade. Needs male and female plants for berry production. Evergreen.
	HALF-RIPE CUTTINGS	Root in a cold frame or sheltered spot outdoors.	September/October	
Azara	HALF-RIPE CUTTINGS	Root in a propagator with slight heat.	August/September	Medium-sized evergreen shrubs grown for foliage and flowers. Mild areas only or frost-free greenhouse elsewhere.
Berberidopsis (Coral Plant)	SEED	Sow in a temperature 16–18°C.	March/April	An evergreen climber up to 4.5 m in height. Shady positions, outdoors in mild areas frost-free greenhouse elsewhere.
	SOFTWOOD CUTTINGS	Root in a propagator at 16–18°C.	April/May	
Berberis (B. vulgaris = Common Barberry)	SEED (may be variable)	Sow in drills in a seedbed outdoors.	Autumn or spring	Large genus of evergreen and deciduous shrubs grown for flowers, fruits and foliage. Species from 30 cm to 3.5 m in height. Some make good hedging plants.
	HALF-RIPE HEEL CUTTINGS	Make cuttings with a heel or section of old wood attached (mallet cutting). Root in a cold frame or propagator.	September/October	
Betula (Birch: B. pendula = Common Silver Birch)	SEED	Sow in pots outdoors to stratify then move to a cold frame or into slight heat in March.	January/February	Medium-sized deciduous trees with ornamental bark for sun or slight shade. Forms can be grafted onto B. pendula.
Buddleia (B. davidii = Butterfly Bush)	SOFTWOOD CUTTINGS	Root in a warm propagator.	March to May	Medium-sized deciduous and evergreen flowering shrubs the evergreen types best left to the mild areas. Can also be raised from seed.
	HARDWOOD CUTTINGS	Root in a cold frame or a sheltered spot outdoors.	November	
Bupleurum	HALF-RIPE CUTTINGS (B. fruticosum)	Root in a propagator at 16–18°C.	July/August	B. fruticosum is a medium-sized evergreen shrub best on chalky soils in mild coastal areas.
Buxus (Box)	SEED	Sow in pots outdoors to stratify then move to a cold frame or slight heat in March.	January/February	Evergreen shrubs with a wide range of species from dwarf shrubs to small trees. Grow in sun or slight shade.
	HALF-RIPE CUTTINGS	Root in a propagator or sheltered spot outdoors.	September/October	
Callicarpa	SEED	Sow in a cold frame or in slight heat.	March/April	Deciduous shrubs mostly grown for their berries and autumn foliage. Species from 1.2 to 3 m in height.
	HALF-RIPE CUTTINGS	Root in a propagator with slight heat.	July/August	

PLANT NAME	METHOD OF PROPAGATION	TREATMENT REQUIRED	TIME OF YEAR PROPAGATED	CULTURAL NOTES
Calluna (C. vulgaris = Heather)	SEED (may benefit from fire or frost action)	Sow in slight heat in a lime-free peat-based compost.	March/April	Small evergreen flowering shrubs for acidic soils in sun or slight shade. 60 cm in height though some varieties are smaller.
	HALF-RIPE CUTTINGS	Root in a cold frame or slightly heated propagator.	July to September	
Calycanthus (C. floridus = Carolina Allspice)	SEED	Sow in a cold frame or in slight heat.	March/April	Deciduous flowering shrubs around 1.8–3 m in height.
	LAYERING	Cut a tongue and peg down appropriate stems.	Late spring/summer	
Camellia	SEED	Sow in slight heat. Seedlings may be variable.	March/April	Medium-sized hardy or slightly tender shrubs, C. japonica being the hardiest. Best in a cold greenhouse or sheltered spot in lime free soils.
	HALF-RIPE or BUD CUTTINGS	Root in a propagator with slight heat.	August	
	LAYERING	Cut a tongue and peg down appropriate shoots.	Spring and autumn	
Campsis (C. radicans = Trumpet Vine)	SEED	Sow in slight heat.	March/April	Deciduous flowering climber for a cold greenhouse or a warm wall in mild areas.
	HALF-RIPE CUTTINGS	Root in a propagator at 16–18°C.	July/August	
	ROOT CUTTINGS	Root in pots in a frost-free greenhouse.	December/January	
Caragana (C. arborescens = Pea Tree)	SEED (soak 24 hours in warm water before sowing)	Sow in a prepared bed outdoors or in a cold frame.	March/April	Mostly spiny deciduous flowering shrubs with species from 60 cm to 6 m in height. Can also be layered in summer.
	HALF-RIPE CUTTINGS	Root in a propagator at 18°C.	July/August	
Carpenteria (C. californica = Californian Mock Orange)	SEED	Sow in a temperature of 16–18°C.	March/April	Evergreen flowering shrubs for mild areas only. 1.8–2.4 m in height. Can also be layered.
	HALF-RIPE CUTTINGS	Root in a warm, moist propagator.	July/August	
Carpinus (Hornbeam)	SEED	Sow in a prepared bed outdoors when ripe or stratify and sow in spring. Use C. betulus as a rootstock for grafting.	In autumn	Small to medium-sized deciduous trees, C. betulus making a good hedge plant. Cultivars are grafted under glass in March.
Carya (Hickory)	SEED	Sow singly in deep pots, stratify and move into slight heat in March. Plant out in autumn disturbing roots as little as possible.	As soon as seed is ripe	Large deciduous trees bearing edible nuts and colouring well in autumn. Dislikes root disturbance.
Caryopteris (Blue Spiraea)	SOFTWOOD CUTTINGS	Root in a warm moist propagator.	April/May	Small deciduous shrubs with blue flowers. Some are rather tender and best in warm spots in mild areas. C. tangutica is the hardiest.
	HALF-RIPE CUTTINGS	Root in a propagator as above. Can also be raised from seed when available.	June to August	
Cassinia	HALF-RIPE CUTTINGS	Root in a cold frame and overwinter there.	September/October	Slightly tender evergreen shrubs up to 1.8 m for peaty soils in mild areas.
Cassiope (syn. Andromeda)	SEED (surface sow)	Sow on peat/sand in slight heat.	March/April	Low-growing evergreen flowering shrubs not more than 30 cm in height. Lime-free soils.
	HALF-RIPE CUTTINGS	Root in a propagator at 18°C.	August	
Castanea (C. sativa = Sweet or Spanish Chestnut)	SEED	Sow in a prepared bed outdoors.	As soon as ripe	Deciduous trees and shrubs bearing edible nuts. Species from small shrubs to large trees.
	WHIP and TONGUE GRAFTING	Graft varietal forms onto C. sativa outdoors.	March	

PLANT NAME	METHOD OF PROPAGATION	TREATMENT REQUIRED	TIME OF YEAR PROPAGATED	CULTURAL NOTES
Catalpa *(C. bignonioides = Indian Bean Tree)*	SEED	Sow in slight heat.	March/April	Medium-sized deciduous trees grown for their flowers and large leaves. Warm positions everywhere.
	HALF-RIPE CUTTINGS	Root short cuttings in a warm propagator.	August	
	ROOT CUTTINGS	Root in pots in a cold frame.	December/January	
Ceanothus	SEED (soak 24 hours in warm water before sowing)	Sow in a temperature of 16–18°C.	March/April	Small to medium-sized deciduous and evergreen shrubs most bearing blue flowers. Grow against a warm wall in cold areas.
	SOFTWOOD CUTTINGS	Root in a propagator at 18°C.	May/June	
	HALF-RIPE CUTTINGS	Root as above, taking cutting with a heel.	August/September	
Cedrus *(Cedar)*	SEED	Sow in pots and place out to stratify. Move to slight heat in March.	As soon as ripe or December/January	Tall wide-spreading coniferous trees best grown in warm sheltered positions. Slow growing. Graft under glass.
	VENEER GRAFTING	Graft varietal forms onto *C. deodara*.	February/March	
Celastrus	SEED	Sow in a prepared bed outdoors.	As soon as ripe or March/April	Deciduous climbers grown for their ornamental fruits. Best grown in groups to ensure fruiting. Some species need greenhouse protection.
	HARDWOOD CUTTINGS	Root in a propagator with slight heat. Can also be layered outdoors in summer.	November/December	
Cephalotaxus *(Plum Yew)*	SEED	Sow in pots, place out to stratify and move into slight heat in March.	As soon as ripe or December/January	Evergreen coniferous shrubs and small trees up to 9 m in height. Plant in sun or slight shade.
	HALF-RIPE CUTTINGS	Root in a cold frame or propagator.	September	
Ceratostigma *(Leadwort; C. plumbaginoides = Plumbago, a herbaceous perennial)*	HALF-RIPE CUTTINGS	Root in a cold frame or propagator.	August	Small deciduous shrubs and perennials bearing blue flowers. They are inclined to be cut to ground level in severe winters.
	DIVISION	Replant outdoors according to the species.	March/April	
Cercidiphyllum	SEED	Sow in a prepared bed outdoors.	When ripe or in spring	Tall deciduous trees grown for their autumn foliage. Rather tender so best grown in mild areas.
	LAYERING	Peg down suitable branches to root.	Spring and summer	
Cercis *(C. siliquastrum = Judas Tree)*	SEED	Sow singly in pots in a temperature of 16–18°C and plant out in final positions as soon as established.	March/April	Medium-sized deciduous trees and shrubs for warm sheltered positions. Grown for their flowers. Varieties are whip and tongue grafted onto species in March outdoors.
Chaenomeles *(Ornamental Quince)*	SEED	Sow in a prepared bed outdoors.	When ripe or in spring	Deciduous early-flowering shrubs most between 90 cm and 3 m in height. Can be grown as a hedge against a south wall.
	LAYERING	Cut a tongue and peg down. Plant out when rooted autumn or spring.	Spring and summer	
Chamaecyparis *(False Cypress)*	SEED (seedlings may be variable)	Sow in pots, place out to stratify and move into slight heat in March.	When ripe or December/January	Evergreen coniferous trees for moist soils in sun or slight shade. *C. lawsoniana* makes a good hedge plant and rootstock for slow-growing varieties.
	HALF-RIPE CUTTINGS	Root in a cold frame or propagator.	September/October	
	VENEER GRAFT	Graft under glass.	March	

PLANT NAME	METHOD OF PROPAGATION	TREATMENT REQUIRED	TIME OF YEAR PROPAGATED	CULTURAL NOTES
Chimonanthus (Winter Sweet)	SEED	Sow in a cold frame or greenhouse.	March/April	Deciduous flowering shrub around 2.4 m in height best grown on a south or south-west wall.
	LAYERING	Cut tongue and peg lower branches down.	Spring and summer	
Chionanthus (Fringe Tree)	SEED	Sow in a cold frame. May benefit from warm/cool stratification.	March/April	Deciduous flowering shrubs or sometimes a small tree 3 m in height. Can also be grafted in spring.
	BUDDING	Bud onto seedling *Fraxinus* (Ash) outdoors.	July/August	
Choisya (Mexican Orange Blossom)	HALF-RIPE CUTTINGS	Root firm cuttings in a cold frame or propagator.	August/September	Small evergreen flowering shrubs for sheltered positions in sun or shade. Best in milder areas.
Cistus (Rock Rose)	SEED	Sow in a cold frame.	March/April	Small to medium-sized evergreen flowering shrubs for warm, dry sunny positions. Most suited to milder areas.
	SOFTWOOD CUTTINGS	Root in a propagator at 18°C.	May/June	
	HALF-RIPE CUTTINGS	Root in a cold frame or greenhouse.	August/September	
Cladrastis (C. lutea = Yellow Wood)	ROOT CUTTINGS	Root in pots in a cold frame.	December/January	Medium-sized deciduous flowering trees. Can also be raised from seed when available.
Clematis (Virgin's Bower)	SEED	Sow in slight heat or a cold frame.	When ripe or in spring	Deciduous and evergreen climbing plants with many species and varieties. Prefer a lime-rich soil in sun or slight shade. X Jackmanii varieties can be cleft grafted onto *C. vitalba* roots under glass in February.
	SERPENTINE LAYERING	Cut tongue and peg shoots down at intervals.	Spring and summer	
	INTER-NODAL HALF-RIPE CUTTINGS	Root in a cold frame or propagator and overwinter in frame.	July/August	
Clerodendrum (C. thomsonae = Glory Bower, an evergreen greenhouse climber)	SEED	Sow in slight heat.	March/April	Large genus of shrubs, small trees and climbers, deciduous and evergreen, all with flowers, only a few hardy in mild areas, the rest needing heat.
	HALF-RIPE CUTTINGS	Root in a propagator at 21°C.	July/August	
	ROOT CUTTINGS	Root hardy types in a cold frame, tender types in heat.	December/January	
Clethra (C. alnifolia = Sweet Pepper Bush)	SEED	Sow in slight heat.	When ripe or in spring	Medium-sized evergreen and deciduous flowering shrubs for lime-free soils. Evergreens best suited to milder areas as they are rather tender.
	HALF-RIPE CUTTINGS	Root in a propagator at 18°C.	July to September	
	DIVISION (offsets: suckers)	Detach suckers where formed and replant.	March	
Colutea (Bladder Senna)	SEED	Sow in slight heat.	March/April	Small to medium-sized deciduous flowering shrubs ideal for hot dry sites.
	HALF-RIPE CUTTINGS	Root short cuttings in a propagator at 18°C.	July to September	
Cornus (Cornel/Dogwood; C. mas = Cornelian Cherry)	SEED	Sow in pots, place out to stratify and move to slight heat in March.	When ripe or December/January	Deciduous trees and shrubs mostly grown for their foliage and ornamental bark though some do flower. Many species some tender for sun or slight shade. Can also be layered.
	DIVISION (suckers)	Separate and replant in moist soils.	March/April	
	HARDWOOD CUTTINGS	Root in a cold frame.	November/December	
Corokia	HALF-RIPE CUTTINGS	Root in a propagator at 18°C. Overwinter in a frost-free position in frames or greenhouse.	August/September	Slightly tender medium-sized evergreen shrubs for mild areas. Flowering and berry-bearing. Can also be layered or raised from seed where available.

PLANT NAME	METHOD OF PROPAGATION	TREATMENT REQUIRED	TIME OF YEAR PROPAGATED	CULTURAL NOTES
Corylopsis	SEED	Sow in a cold frame or in slight heat. May benefit from warm/cool stratification.	March/April	Medium-sized deciduous flowering shrubs and small trees. Grow in sheltered positions in sun or slight shade. Can also be layered.
	SOFTWOOD CUTTINGS	Root in a propagator at 18°C.	April to June	
Cotinus (C. coggygria, syn. Rhus cotinus = Smoke Tree)	SEED	Sow in a cold frame.	March/April	Medium to large deciduous shrubs, many grown for their autumn foliage. Can also be layered in summer.
	HALF-RIPE CUTTINGS	Root in a warm propagator and grow on overwinter in a cold frame.	August/September	
Cotoneaster	SEED	Sow in prepared beds outdoors or stratify overwinter in peat/sand.	When ripe or in spring	Mostly deciduous and evergreen shrubs with a range of sizes to suit most situations. Mainly grown for their colourful berries.
	HALF-RIPE CUTTINGS	Root in a propagator at 18°C.	August/September	
Crataegus (Thorn; C. monogyna = Hawthorn)	SEED (slow to germinate)	Sow in pots or outdoor beds after winter stratification.	March/April	Deciduous flowering trees and medium shrubs, C. monogyna being the hedging type. Can also be air-layered in summer or whip and tongue grafted in March.
	BUDDING	Bud varieties onto C. monogyna 5–8 cm above soil level.	June to August	
Crinodendron (syn. Tricuspidaria; C. hookeranum = Lantern Tree)	HALF-RIPE CUTTINGS	Root in a warm shady propagator and grow on under glass.	August	Medium to large evergreen flowering shrubs for milder areas or in frost-free greenhouses elsewhere. Plant in slight shade.
	SEED	Sow in slight heat and grow on in frost-free positions.	When ripe or in spring	
Cryptomeria (C. japonica = Japanese Cedar)	SEED	Sow in a cold frame in autumn or a bed outdoors in spring.	As soon as ripe	Tall evergreen conifer for sheltered positions in moist soils. Many varieties much smaller than type. Can be veneer grafted onto the true species.
	HALF-RIPE CUTTINGS	Root in a cold frame or propagator.	September/October	
Cupressocyparis (C. leylandii is a hybrid between Cupressus macrocarpa and Chamaecyparis nootkatensis)	HALF-RIPE CUTTINGS	Root in a cold frame or propagator.	September/October	Fast-growing conifers. Evergreen and forming large trees. Plant in sun or slight shade. Good hedge or screen plants.
Cupressus (Cypress)	SEED	Sow in a cold frame.	March/April	Medium-sized evergreen coniferous trees for sheltered positions and mild areas only for some kinds. Those difficult to root as cuttings can be grafted.
	HALF-RIPE CUTTINGS	Root in a warm propagator. Grow on in slight heat.	September/October	
	VENEER GRAFTING	Graft under glass onto C. macrocarpa.	March	
Cytisus (Broom; C. proliferus = Lucerne Tree, a greenhouse shrub)	SEED (some may benefit from fire action)	Sow hardy types in a cold frame or outdoors, tender types in slight heat.	March/April	Deciduous and evergreen small to medium-sized flowering shrubs and small trees growing well on dry stony soils. Some species require greenhouse cultivation.
	HALF-RIPE HEEL CUTTINGS	Root 8 cm cuttings in a warm propagator.	August/September	
Daboecia (D. cantabrica = St. Daboec's Heath)	SEED	Sow in slight heat on a lime-free peat compost.	March/April	Small evergreen flowering shrubs for lime-free soils. Best suited to milder areas as they are slightly tender.
	HALF-RIPE CUTTINGS	Root in a cold frame or propagator.	June to September	
Danae (D. racemosa = Alexandrian Laurel)	SEED	Sow in a prepared bed outdoors.	In autumn or spring	Evergreen shrub up to 1.2 m in height. Foliage is good for cutting.
	DIVISION	Replant in moist soil in slight shade.	March/April	

PLANT NAME	METHOD OF PROPAGATION	TREATMENT REQUIRED	TIME OF YEAR PROPAGATED	CULTURAL NOTES
Daphne	SEED	Sow in pots, place out to stratify and move to slight heat in March.	When ripe or December/January	Mostly small evergreen or deciduous flowering shrubs ideal for rock gardens especially on lime-free soils. Can also be layered in summer and autumn.
	HALF-RIPE CUTTINGS	Root in a propagator at 18°C.	August/September	
Davidia (D. involucrata = Handkerchief Tree)	SEED	Sow in a cold frame.	March/April	A deciduous tree up to 18 m in height grown for the bracts surrounding the flowers.
	HALF-RIPE CUTTINGS	Root in a propagator at 18°C.	August/September	
	AIR LAYERING	Layer as described on page 29.	During summer	
Decaisnea	SEED	Sow in slight heat and grow on in a cold frame. Cuttings will also root.	March/April	Medium-sized deciduous shrub grown for its ornamental edible fruits.
Desfontainea	SEED	Sow in a temperature of 16–18°C.	March/April	Evergreen flowering shrubs up to 3 m. Best in a sheltered spot in mild areas or frost-free greenhouse. Can also be layered.
	HALF-RIPE TIP CUTTINGS	Root in a warm propagator.	September/October	
Deutzia	HALF-RIPE CUTTINGS	Root in a cold frame or propagator.	June to August	Deciduous flowering shrubs, most less than 2.4 m in height. Can also be raised from seed when available.
	HARDWOOD CUTTINGS	Root 25–30 cm cuttings in a cold frame.	November/December	
Diervilla (Honeysuckle Bush)	HALF-RIPE CUTTINGS	Root in a cold frame or propagator.	June to August	Small deciduous flowering shrubs for moist soil in sun or slight shade. D. lonicera also produces suckers.
	HARDWOOD CUTTINGS	Root in a cold frame.	November/December	
Dipelta	HALF-RIPE CUTTINGS	Root in a warm propagator and overwinter under glass.	August/September	Deciduous flowering shrubs similar to Weigela. Can also be raised from seed.
Distylium	HALF-RIPE CUTTINGS	Root in a warm moist propagator and overwinter under glass.	August/September	D. racemosum is a small evergreen flowering shrub best grown in mild areas.
Drimys (D. colorata = Pepper Tree)	HALF-RIPE CUTTINGS	Root in a cold frame and overwinter there. Can also be layered in spring.	August/September	Medium-sized evergreen shrubs or small trees grown for foliage and flowers for milder areas only.
Elaeagnus	SEED	Sow in a prepared bed outdoors.	In autumn	Medium-sized deciduous and evergreen shrubs grown for foliage and/or flowers. Can also be layered in spring.
	HALF-RIPE CUTTINGS	Root in a warm propagator.	September/October	
Embothrium (E. coccineum = Chilean Fire Bush)	SEED	Sow in sand/peat in a temperature of 18–21°C.	March/April	Medium to large evergreen flowering shrubs or small trees for mild areas and lime-free soils.
	ROOT CUTTINGS	Root in slight heat and plant out after hardening off in late spring.	December/January	
Enkianthus	SEED	Sow in slight heat in a lime-free peat compost.	March/April	Medium-sized deciduous shrubs for lime-free soils in sun or slight shade. Grown for their flowers and autumn foliage.
	HALF-RIPE CUTTINGS	Root in a warm propagator. Can also be layered in summer.	August/September	
Erica (Heath/Heather)	SEED (some may benefit from fire action)	Sow on the surface of peat/sand in slight heat. Grow on in a shady frame.	March/April	A large genus of evergreen flowering shrubs, most hardy species requiring lime-free soils in sun or slight shade. Cape heaths need cool but frost-free conditions all year.
	HALF-RIPE CUTTINGS	Root 2.5 cm cuttings in a cold frame or propagator.	June to September	
Cape Heaths	TIP CUTTINGS	Root in a peat-based compost in a propagator at 16°C.	November	

PLANT NAME	METHOD OF PROPAGATION	TREATMENT REQUIRED	TIME OF YEAR PROPAGATED	CULTURAL NOTES
Eriobotrya *(Loquat)*	SEED HALF-RIPE CUTTINGS	Sow in slight heat. Root in a warm propagator.	March/April August/September	Tall evergreen flowering shrub with large leaves and edible fruits for sheltered positions in mild areas.
Erythrinia *(E. crista-galli = Coral Tree)*	SEED BASAL HEEL CUTTINGS	Sow in a temperature of 18–21°C. Root in a propagator at 18–21°C.	March/April In spring	Deciduous or semi-evergreen medium-sized shrubs and trees most suited to frost-free greenhouses.
Escallonia	SEED HALF-RIPE CUTTINGS	Sow in a temperature of 16–18°C. Root in a propagator in slight heat.	March/April July to October	Mostly medium-sized evergreen flowering shrubs and small trees best grown in mild areas especially near the sea.
Eucalyptus *(Gum Tree)*	SEED	Sow in a temperature of 18–24°C. Some species' seed may benefit from fire or frost action to assist germination.	March/April	Medium to tall evergreen flowering/foliage trees with many species, some hardy and others needing a cool greenhouse.
Eucryphia	SEED HALF-RIPE CUTTINGS AIR LAYERING	Sow in slight heat in a peat-based lime-free compost. Root in a warm propagator and grow on under glass. Layer as described on page 29.	March/April June/July During summer	Small to medium-sized evergreen and deciduous trees, some slightly tender. *E. glutinosa* is the hardiest. Plant in moist lime-free soils.
Euonymus *(Spindle Tree)* *E. fortunei (syn. radicans)*	SEED HALF-RIPE CUTTINGS (evergreens) DIVISION	Sow in slight heat or a prepared bed outdoors. Root in a warm propagator. Replant outdoors.	March/April July or September/October October	Medium-sized evergreen and deciduous shrubs and small trees for sun or slight shade. Mostly grown for their foliage and/or ornamental fruits.
Euryops	SOFTWOOD CUTTINGS	Root in a warm propagator and overwinter under glass. Can also be raised from seed.	April to June	Small evergreen flowering shrubs most requiring frost-free greenhouses but some are hardy in mild areas.
Exochorda *(Pearl Bush)*	SEED SOFTWOOD CUTTINGS	Sow in a cold frame or greenhouse. Root in a warm propagator.	In autumn or March/April May/June	Medium-sized deciduous flowering shrubs which can also be layered in summer.
Fabiana *(False Heath)*	HALF-RIPE CUTTINGS	Root in a warm propagator and overwinter under glass.	August	Small evergreen flowering shrubs best in sheltered spots in mild areas or a frost-free greenhouse elsewhere.
Fagus *(Beech)*	SEED WHIP and TONGUE GRAFT	Stratify in cool conditions through winter then sow outdoors in a seedbed. Graft selected types onto seedlings outdoors.	March/April March	Medium to large deciduous trees growing well on lime-rich soils. *F. sylvatica* is used as a hedge plant and as a rootstock.
Fitzroya	HALF-RIPE CUTTINGS	Root in a warm propagator and overwinter under glass.	August	Medium-sized evergreen coniferous shrubs for sheltered positions. Can be raised from seed when available.
Forsythia *(Golden Bells)*	HALF-RIPE CUTTINGS HARDWOOD CUTTINGS	Root 13 cm cuttings in a warm propagator. Root in a cold frame or sheltered spot outdoors.	June/July November/December	Small to medium-sized deciduous flowering shrubs for sun or slight shade. Can also be layered in summer and autumn.

PLANT NAME	METHOD OF PROPAGATION	TREATMENT REQUIRED	TIME OF YEAR PROPAGATED	CULTURAL NOTES
Fothergilla *(American Witch Hazel)*	SEED (slow germination)	Sow in a temperature of 16–18°C.	March/April	Deciduous flowering shrubs with good autumn foliage. Best in lime-free soils in sun or slight shade. Can also be layered.
	HALF-RIPE CUTTINGS	Root in a warm propagator.	July	
Fraxinus *(Ash)*	SEED	Sow in a seedbed outdoors and transplant when 1 year old.	In autumn or March/April	Medium to large deciduous trees grown for their foliage and sometimes flowers. Grow in sun or slight shade.
	WHIP and TONGUE GRAFT	Graft forms outdoors onto species.	March	
Fremontodendron *(syn. Fremontia)*	SEED	Sow in slight heat and grow on in pots.	February/March	Medium-sized flowering evergreen shrubs or small trees best in sheltered positions in mild areas.
	SOFTWOOD CUTTINGS	Root in a warm propagator.	May/June	
Garrya *(G. elliptica = Silk Tassel Bush)*	HALF-RIPE CUTTINGS	Root in a warm propagator, overwinter under glass and grow on in pots until planted out.	August	Medium-sized evergreen catkin-bearing shrubs with male and female plants. Sun or slight shade in the protection of a wall in cold areas.
Gaultheria	SEED	Sow in a peaty compost outdoors or in slight heat March/April.	In autumn	Mostly small evergreen shrubs grown for foliage flowers and berry-like fruits. Grow best on lime-free soils in slight shade.
	HALF-RIPE CUTTINGS	Root short cuttings in a warm propagator.	August/September	
	DIVISION (offsets)	Separate offsets when available and replant.	March/April	
Genista *(Broom and Gorse; G. aetnensis = Mt. Etna Broom; G. hispanica = Spanish Gorse)*	SEED (some may benefit from fire action)	Sow in slight heat or in a cold frame a little later.	February/March	Dwarf to medium-sized deciduous flowering shrubs for well-drained positions. Grow on in pots as they resent root disturbance.
	HALF-RIPE CUTTINGS	Root cuttings with a heel in a cold frame or propagator.	September/October	
Ginkgo *(G. biloba = Maidenhair Tree)*	SEED	Sow in a cold frame or seedbed outdoors.	March/April	A medium to large deciduous coniferous tree mostly grown for its foliage. Varieties can be grafted onto seedlings under glass.
	AIR LAYERING	Layer as described on page 29.	April to August	
Gleditschia *(G. triacanthos = Honey Locust)*	SEED	Carefully chip the seed and sow in slight heat.	February/March	Medium to large spiny deciduous trees mostly grown for their foliage. Good for planting in towns.
	WHIP AND TONGUE GRAFT	Graft varieties onto seedling species outdoors.	March	
Griselinia	HALF-RIPE CUTTINGS	Root in a cold frame or propagator. Can also be raised from seed when available.	September/October	Medium-sized evergreen shrubs grown for their foliage. Only suited to mild areas especially near the sea.
Gymnocladus *(G. dioicus, syn. canadensis = Kentucky Coffee Tree)*	SEEDS (soak 24 hours in warm water before sowing)	Soak or chip seeds then sow in slight heat.	February/March	Medium-sized deciduous trees grown mostly for their large ornamental leaves. Grows very slowly. *G. chinesis* best in mild areas.
	ROOT CUTTINGS	Root in a warm spot under glass.	December/January	
Halesia *(H. carolina = Silver Bell/Snowdrop Tree)*	SEED	Sow in a cold frame or greenhouse.	March/April	Small deciduous flowering trees for sheltered positions in sun or slight shade and lime-free soils.
	SOFTWOOD CUTTINGS	Root in a warm propagator. Can also be layered in summer.	May/June	
Halimium	SEED	Sow in a temperature of 16–18°C.	March/April	Small evergreen flowering shrubs best suited to mild areas.
	HALF-RIPE CUTTINGS	Root in a cold frame and overwinter there.	July/August	

PLANT NAME	METHOD OF PROPAGATION	TREATMENT REQUIRED	TIME OF YEAR PROPAGATED	CULTURAL NOTES
Hamamelis (Witch Hazel)	SEED (slow germination)	Sow in a cold frame or greenhouse. May benefit from warm/cool stratification.	March/April	Medium-sized deciduous flowering shrubs or small trees. Prefer lime-free soils in sun or slight shade. Rare types can be side grafted and then grown in a propagator until a union is made.
	AIR LAYERING	Air layer outdoors as described on page 29.	Spring and summer	
	SIDE GRAFT	Graft rare types onto *H. virginiana* 7 cm above compost surface.	March or August	
Hebe (Shrubby Veronica)	SOFTWOOD CUTTINGS	Root in a warm propagator.	April/May	Small to medium-sized flowering evergreen shrubs most tender and best in mild coastal areas or a frost-free greenhouse elsewhere.
	HALF-RIPE CUTTINGS	Root in a cold frame and overwinter there. Can also be raised from seed if available.	July to September	
Hedysarum (H. coronarium = French Honeysuckle, a herbaceous perennial)	SEED	Sow in a temperature of 16–18°C.	March/April	*H. multijugum* is a deciduous flowering shrub up to 1.5 m. The genus also contains a number of herbaceous perennials and sub-shrubs
	LAYERING	Cut a tongue and peg appropriate shoots down.	In spring	
Hibiscus (H. syriacus = Bush Mallow)	SEED	Sow in slight heat.	March/April	*H. syriacus* is a hardy deciduous shrub up to 3 m in height. There are many varieties and other species for greenhouse cultivation.
	HALF-RIPE CUTTINGS	Root in a warm propagator.	August/September	
	LAYERING	Cut a tongue and peg down appropriate shoots.	During summer	
Hippophae (Sea Buckthorn)	SEED	Stratify in pots through winter then sow in a seedbed outdoors.	In autumn or December/January	Large deciduous berry-bearing shrubs or small trees, both male and female plants being needed for berry production. Good in exposed areas.
	LAYERING	Cut a tongue and peg down. Suckers can also be used where they form.	October/November	
Hoheria (Ribbon Wood)	HALF-RIPE CUTTINGS	Root in a warm propagator.	August/September	Large evergreen and deciduous flowering shrubs or small trees best suited to mild areas. Also by seed if available sown in heat.
	LAYERING	Cut a tongue and peg appropriate shoots down to root.	April to August	
Holodiscus	SEED	Sow in a cold frame.	March/April	Small to medium-sized deciduous flowering shrubs for sun or slight shade.
	HALF-RIPE CUTTINGS	Root in a cold frame or propagator.	August/September	
	HARDWOOD CUTTINGS	Root in a cold frame.	November	
Hydrangea	SEED	Sow in slight heat.	March/April	Medium-sized deciduous flowering shrubs and vigorous climbers, some species rather tender.
	HALF-RIPE CUTTINGS	Root 10 cm cuttings in a warm propagator.	July/August	
H. petiolaris (Climbing Hydrangea)	SERPENTINE LAYERING	Cut tongues and peg down at intervals.	During spring and summer	*H. petiolaris* is a self-clinging climber ideal for north walls.
Hypericum (St. John's Wort; H. calycinum = Rose of Sharon)	SEED	Sow in slight heat.	March/April	Flowering evergreen and deciduous small shrubs, sub-shrubs and perennial plants some slightly tender. Full sun except *H. calycinum* which will grow in shade making a good ground cover.
	HALF-RIPE CUTTINGS	Root in a cold frame or propagator.	September/October	
	DIVISION	Divide where possible and replant according to the species.	March/April	
Idesia	SEED	Sow in a temperature of 18–21°C.	March/April	Deciduous flowering and berry-bearing tree up to 12 m in height.
	AIR LAYERING	Layer outside as described on page 29.	During spring	

PLANT NAME	METHOD OF PROPAGATION	TREATMENT REQUIRED	TIME OF YEAR PROPAGATED	CULTURAL NOTES
Ilex (Holly; 1. aquifolium = Common Holly; many varieties and forms)	SEED	Stratify for 1 year in peat/sand then sow outdoors or in a cold frame.	As soon as ripe	Evergreen and deciduous medium to large shrubs and medium-sized trees some rather tender. Berry-bearing and foliage plants with male and female forms. Can also be layered in summer.
	HALF-RIPE CUTTINGS	Root in a propagator at 18°C.	August/September	
	BUDDING	Bud onto seedling species outdoors or graft in March.	May	
Itea	HALF-RIPE HEEL CUTTINGS	Root in a propagator with slight heat.	August/September	Small to medium-sized evergreen and deciduous flowering shrubs for sun or slight shade. Can also be layered in spring and summer.
	DIVISION (suckers)	Separate where they form and replant.	In spring	
Jasminum (Jasmine; J. nudiflorum = Winter Jasmine)	HALF-RIPE HEEL CUTTINGS	Root and overwinter in a cold frame.	July/August	A varied genus of evergreen and deciduous climbing, trailing and shrubby plants. The propagation here is for hardy types; tender kinds need more heat.
	HARDWOOD HEEL CUTTINGS	Root as above.	November	
	SERPENTINE LAYERING	Cut tongues and peg down in pots.	During spring and summer	
Juniperus (Juniper; J. communis bears edible berries, or cones, used as seasoning)	SEED	Stratify berries (cones) then sow in a seedbed outdoors in March/April.	As soon as ripe	Evergreen coniferous trees with many species varieties and forms from dwarf prostrate plants to tall columns. They can also be layered outdoors in spring and summer. Use J. virginiana as the rootstock for J. chinensis.
	HALF-RIPE CUTTINGS	Root short cuttings in a cold frame or propagator.	September to January	
	VENEER GRAFT	Graft forms onto a suitable rootstock under glass.	March	
Kalmia (K. latifolia = Calico Bush/American Laurel)	SEED	Stratify overwinter then sow in slight heat.	March/April	Small evergreen flowering shrubs for lime-free soils and slight shade.
	HALF-RIPE CUTTINGS	Root in sand/peat in a warm propagator.	August/September	
Kalmiopsis	HALF-RIPE CUTTINGS	Root in sand/peat in a warm propagator. Can also be raised from seed.	July to September	Dwarf evergreen flowering shrub for lime-free soils.
Kerria (Jew's Mallow)	HALF-RIPE CUTTINGS	Root 5–8 cm cuttings in a cold frame.	June/July	Small deciduous flowering shrub for sun or slight shade. Suitable branches can layered in summer.
	DIVISION (suckers)	Separate and replant.	March	
Koelreuteria (K. paniculata = Golden Rain Tree)	SEED (chip before sowing)	Sow in a temperature of 16–18°C.	March/April	Small deciduous trees ornamental in all aspects. K. paniculata and K. apiculata are hardy the rest tender.
	ROOT CUTTINGS	Root in slight heat under glass.	December/January	
Kolkwitzia (K. amabilis = Beauty Bush)	HALF-RIPE CUTTINGS	Root in a sandy compost in a cold frame or propagator.	July/August	Deciduous flowering shrub up to 3 m in height. Can also be raised from seed.
Laburnocytisus	WHIP and TONGUE GRAFTING	Graft onto seedling Laburnum anagyroides outdoors.	February/March	Small deciduous flowering tree, a bi-generic graft hybrid.
Laburnum	SEED	Soak in warm water till cool before sowing in a cold frame.	March/April	Small to medium-sized deciduous flowering trees bearing poisonous seeds. Varieties can be budded onto the common species in July.
	HALF-RIPE CUTTINGS	Root in a propagator at 18°C.	July/August	
Larix (Larch)	SEED	Stratify over winter then sow outdoors or in a cold frame March/April.	As soon as ripe	Tall deciduous coniferous trees. Varieties can be grafted onto L. decidua.
	VENEER GRAFT	Graft in slight heat.	March	

PLANT NAME	METHOD OF PROPAGATION	TREATMENT REQUIRED	TIME OF YEAR PROPAGATED	CULTURAL NOTES
Ledum (L. palustre = Wild Rosemary)	SEED	Sow in slight heat.	March/April	Small evergreen flowering shrubs for lime-free soils. Can also be layered in summer outdoors.
	HALF-RIPE CUTTINGS	Root in a propagator at 18°C.	July/August	
	DIVISION	Replant according to species.	March/April	
Leiophyllum (L. buxifolium = Sand Myrtle)	HALF-RIPE CUTTINGS	Root in a propagator at 18°C. Can also be raised from seeds and layers.	July/August	Evergreen flowering shrub up to 45 cm in height for lime-free soils.
Leptospermum (L. scoparium = South Sea Myrtle)	SEED	Sow in sand/peat at 16–18°C.	March/April˙	Small to medium-sized evergreen flowering shrubs or small trees for mild areas in lime-free well-drained soils.
	HALF-RIPE CUTTINGS	Root in a propagator at 18°C.	August	
Lespedeza (Bush Clover)	SEED (warm water soak till cool)	Sow in slight heat or cold frame.	March/April	Small, flowering shrubs which tend to be herbaceous in cold areas sending up annual shoots. Division can be used where this occurs.
	BASAL CUTTINGS	Root in a warm propagator. Can also be raised from cuttings in summer.	In spring	
Leycesteria (L. formosa = Himalayan Honeysuckle)	SEED	Sow in slight heat.	March/April	Medium-sized deciduous flowering shrubs best in frost-free greenhouses in colder areas.
	HALF-RIPE CUTTINGS	Root in a cold frame or propagator.	August/September	
	HARDWOOD CUTTINGS	Root in a cold frame.	November/December	
Libocedrus (syn. Calocedrus; L. decurrens = Incense Cedar)	SEED	Stratify outdoors over winter then sow in a cold frame March/April.	As soon as ripe	Tall evergreen coniferous trees, some rather tender. L. decurrens is the hardiest.
	HALF-RIPE CUTTINGS	Root in a cold frame or in slight heat.	September/October	
Ligustrum (Privet)	SEED	Stratify outdoors over winter then sow outdoors or in a cold frame March/April.	As soon as ripe	Medium to tall deciduous and evergreen shrubs, most grown for their foliage though some have good flowers. L. ovalifolium is used for hedging.
	HALF-RIPE CUTTINGS	Root in a cold frame or heated propagator.	July/August	
L. ovalifolium and L. vulgare	HARDWOOD CUTTINGS	Root outdoors or in a cold frame.	November/December	
Lindera (L. benzoin = Spice Bush, probably the hardiest species)	SEED	Stratify over winter then sow in a cold frame March/April.	As soon as ripe	Medium to tall evergreen and deciduous flowering shrubs and trees, most being rather tender. Suit lime-free soils in sun or slight shade.
	HALF-RIPE CUTTINGS	Root in a propagator at 18°C.	July/August	
Liquidambar (L. styraciflua = Sweet Gum)	SEED	Stratify at 4°C over winter then sow outdoors or in a cold frame. Protect from frost in their first few years.	March/April	Medium-sized deciduous trees best in moist soils and sheltered positions in milder areas. Can also be air layered in summer.
Liriodendron (Tulip Tree)	SEED	Place out to stratify then sow in a cold frame or outdoors in prepared seedbed.	As soon as ripe	Large deciduous flowering trees for deep rich soils. Varieties can be grafted onto L. tulipifera under glass in spring.
Loiseleuria (L. procumbens = Alpine Azalea)	SEED	Sow in slight heat.	March/April	Mat-forming dwarf evergreen flowering shrub for well-drained exposed situations.
	DIVISION (offsets)	Divide and replant in lime-free soils. Can also be raised from cuttings.	In autumn	
Lonicera (Honeysuckle)	SEED	Place outdoors to stratify then sow in slight heat March/April.	As soon as ripe	Varied genera of deciduous and evergreen hardy and tender shrubs and climbers. L. nitida can be used as a low hedging plant. Grow in sun or slight shade.
	HALF-RIPE CUTTINGS	Root in a propagator at 16–18°C. Can also be layered outdoors in summer.	June/July	

PLANT NAME	METHOD OF PROPAGATION	TREATMENT REQUIRED	TIME OF YEAR PROPAGATED	CULTURAL NOTES
Loropetalum	SEED	Sow in a temperature of 16–18°C.	March/April	Medium-sized evergreen flowering shrub for a cool greenhouse or outdoors in very mild areas.
	HALF-RIPE CUTTINGS	Root in a propagator at 18°C. Can also be layered.	July/August	
Lycium (Box Thorn)	SEED	Sow in slight heat.	March/April	Medium-sized deciduous shrubs, some inclined to ramble. Grown mainly for their berries *L. chinense* and *L. halimifolium* grow well by the sea.
	HALF-RIPE CUTTINGS	Root in a propagator at 16–18°C.	July/August	
	DIVISION (suckers)	Separate and replant in sandy well-drained soils.	March/April	
Maackia	SEED	Sow in slight heat.	March/April	Usually medium-sized deciduous flowering shrubs, sometimes forming small trees.
	ROOT CUTTINGS	Root and establish in slight heat.	December/January	
Magnolia	SEED	Sow in slight heat.	Autumn or spring	Deciduous and evergreen flowering trees and shrubs, some rather tender. Grow in sheltered positions out of spring frost-pockets in sun or slight shade preferably in lime-free soils.
	AIR LAYERING	Layer as described on page 29. Can also be layered in the ordinary way.	Spring and summer	
M. grandiflora	SOFTWOOD CUTTINGS	Root in a warm propagator.	April/May	
Mahonia (*M. aquifolium* = Oregon Grape)	SEED	Stratify at 4°C over winter then sow in slight heat March/April.	As soon as ripe	Evergreen flowering shrubs, some a little tender. *M. aquifolium* is hardy and can be grown in sun or in the shade of trees and can sometimes be divided.
	BUD CUTTINGS	Root in a propagator at 18°C.	April to October	
Malus (species; for cultivars, budding and grafting, see Apple, Chapter 13)	SEED	Stratify outdoors then sow outdoors or in pans in a cold frame March/April.	As soon as ripe	Deciduous flowering small trees and shrubs bearing ornamental and sometimes edible fruits. Good autumn foliage.
Margyricarpus (Pearl Fruit)	SEED	Stratify over winter then sow in a cold frame March/April.	As soon as ripe	*M. setosus* is a dwarf evergreen for the rock garden mainly grown for its white berries.
	HALF-RIPE CUTTINGS	Root in a peaty compost in a warm propagator.	July/August	
Melia (*M. azedarach* = Bead Tree/Indian Lilac)	SEED	Sow in a temperature of 18–21°C. Can also be raised from summer cuttings.	March/April	Medium-sized deciduous flowering trees best confined to very mild areas or cool greenhouse.
Menziesia	SEED	Sow in sand/peat in a temperature of 16–18°C.	March/April	Small deciduous flowering shrubs for lime-free soils in sun or slight shade. Can also be layered.
	HALF-RIPE CUTTINGS	Root in a propagator at 18°C.	July/August	
Metasequoia (*M. glyptostroboides* = Dawn Redwood)	SEED	Sow in a cold frame or greenhouse.	When ripe or March/April	Tall deciduous coniferous tree with good autumn foliage. One of *the* living fossils. Grow in rich moist soils.
	HARDWOOD CUTTINGS	Root 15 cm cuttings singly in small pots in a propagator at 18°C.	November to January	
Metrosideros	SOFTWOOD CUTTINGS	Root in a propagator at 16–18°C. Grow on in neutral or lime-free soils.	May to July	Small to medium-sized evergreen trees for very mild areas or cool greenhouses.
Muehlenbeckia (*M. complexa* = Wire Vine)	HALF-RIPE CUTTINGS	Root in a cold frame.	July/August or October/November	Creeping or climbing deciduous shrubs best in mild sheltered gardens, *M. axillaris* making good ground cover.
	DIVISION	Replant according to the species. Some are greenhouse plants.	March	

PLANT NAME	METHOD OF PROPAGATION	TREATMENT REQUIRED	TIME OF YEAR PROPAGATED	CULTURAL NOTES
Mutisia (Climbing Gazania)	SEED HALF-RIPE CUTTINGS	Sow in sand/peat at 16–18°C. Root in a propagator at 18°C.	March/April July/August	Flowering tendril climbers some greenhouse plants but others such as *M. decurrens* are hardy in mild areas.
Myrica (*M. cerifera* = Wax Myrtle. *M. gale* = Bog Myrtle)	SEED DIVISION	Wash in warm water and place out to stratify. Sow in sand/peat in a cold frame. Replant according to the species or layer in summer.	As soon as ripe March/April March/April	Small to medium-sized deciduous and evergreen shrubs grown for their catkins and berries. Grow in lime-free moist soils.
Myrtus (Myrtle)	SEED HALF-RIPE CUTTINGS	Sow in a temperature of 16–18°C. Root in a propagator at 18°C.	March/April July/August	Medium-sized evergreen shrubs with fragrant leaves and flowers. Grow in warm positions or in a frost-free greenhouse in cold areas.
Nandina (*N. domestica* = Heavenly/Sacred Bamboo)	SEED HALF-RIPE CUTTINGS	Sow in a temperature of 16–18°C. Root in a propagator at 16–18°C.	As soon as ripe July/August	Bamboo-like evergreen flowering shrub with red berries. Grow in warm sheltered spots in mild areas.
Neillia	HALF-RIPE CUTTINGS	Root in a propagator at 18°C.	July/August	Small deciduous flowering shrubs. Can also be grown from suckers.
Nothofagus (Southern Beech)	SEED HALF-RIPE CUTTINGS AIR LAYERING	Sow outdoors or in a cold frame. Root in a warm propagator. Layer outdoors as described on page 29.	When ripe or March/April July/August Spring and summer	Medium to large evergreen and deciduous trees, most only hardy in the mildest areas. *N. obliqua* and *N. procera* are the hardiest. Plant on lime-free soils in sun or slight shade.
Notospartium (Southern Broom)	SEED	Sow in a cold frame or greenhouse and grow on singly in small pots until planted out.	March/April	Evergreen flowering shrubs similar to broom best grown in milder areas. Can also be raised from cuttings.
Nyssa (Tupelo Tree)	SEED AIR LAYERING	Stratify outdoors over winter then sow singly in small pots in slight heat March/April. Layer outdoors as described on page 29.	As soon as ripe/January/February Spring and summer	Medium to large deciduous trees grown for their autumn foliage. Plant out carefully to avoid root disturbance.
Olearia (Daisy Bush; *O. macrodonta* = New Zealand Holly)	HALF-RIPE CUTTINGS	Root in a sandy compost in a cold frame or propagator with slight heat.	August/September	Small to medium-sized evergreen flowering shrubs *O. haastii* being the hardiest and the rest more suited to coastal areas or frost-free greenhouses.
Osmanthus (*O. fragrans* = Fragrant Olive; greenhouse most areas)	SEED HALF-RIPE CUTTINGS	Sow in sand/peat at 16–18°C. Root in a propagator with slight heat.	March to May September/October	Mostly medium to large evergreen flowering shrubs for sheltered positions. Can also be layered in summer.
Osmarea (Bi-generic hybrid; Osmanthus x phillyrea)	HALF-RIPE CUTTINGS	Root in a cold frame or propagator.	September/October	Evergreen flowering shrub making good hedges or screens in sun or slight shade on dry sites.
Osmaronia (Indian Plum)	DIVISION (suckers)	Separate and replant in sun or slight shade.	November to March	Deciduous fragrant shrub up to 1.5 m in height. Can also be raised from seed.
Osteomeles	SEED HALF-RIPE CUTTINGS	Sow in a temperature of 16–18°C. Root in a propagator at 18°C.	March/April July/August	Small evergreen flowering shrubs for lime-free soils. Best in mild areas or against a warm wall elsewhere.

PLANT NAME	METHOD OF PROPAGATION	TREATMENT REQUIRED	TIME OF YEAR PROPAGATED	CULTURAL NOTES
Ostrya (Hop Hornbeam)	SEED	Place out to stratify then sow in a seedbed outdoors March/April.	As soon as ripe	Small to medium-sized deciduous trees with ornamental fruits.
Oxydendrum (O. arboreum = Sorrel Tree)	SEED	Sow in slight heat in a sand/peat compost.	When ripe or March/April	Small deciduous tree or tall shrub grown for its flowers and autumn foliage. Grow in lime-free soils protected from cold winds.
	HALF-RIPE CUTTINGS	Root in a propagator at 16–18°C. Can also be layered outdoors in summer.	August/September	
Paeonia (Tree Paeony; P. suffruticosa = Moutan Paeony)	SEED LAYERING	Sow in slight heat. Cut tongue and peg appropriate shoots down outdoors.	When ripe March/April	Wide-spreading deciduous flowering shrubs for rich, deep soils in sun or slight shade. Need protection from late spring frosts. Pot grafts so that junction is below compost level and plant out so that it is 8 cm below ground level in spring.
	WEDGE GRAFTING	Graft non-flowering shoots onto the roots of P. lactiflora and P. officinalis. Keep in a cold frame over winter.	July/August	
Parrotia (Iron Tree)	SEED	Sow in a cold frame. May benefit from warm/cool stratification. Can also be layered outdoors in summer.	As soon as ripe	Small to medium-sized deciduous trees grown for their flowers and autumn foliage.
Parthenocissus (P. quinquefolia = True Virginia Creeper)	SEED	Sow in a cold frame or seedbed outdoors.	March/April	Mostly hardy deciduous climbers grown for their autumn foliage. Grow in sun or slight shade self-clinging types ideal for north walls.
	SOFTWOOD CUTTINGS	Root in a propagator at 18°C.	April/May	
	HARDWOOD CUTTINGS	Root as above.	November/December	
	SERPENTINE LAYERING	Peg appropriate stems down into pots.	Spring and summer	
Paulownia	SEED	Sow in a cold frame or in slight heat.	When ripe or March/April	Medium-sized deciduous ornamental flowering trees for warm sheltered positions. Good foliage plant topped back yearly.
	ROOT CUTTINGS	Root in sandy compost in cold frame or frost-free greenhouse.	December/January	
Periploca (P. graeca = Silk Vine)	HALF-RIPE CUTTINGS	Root in a propagator at 18°C.	July/August	Deciduous flowering climbing shrubs also grown for foliage and fruits. Twining so suitable for trellises etc.
	DIVISION	Replant in suitable positions. Can also be layered outdoors in summer.	March/April	
Pernettya (Prickly Heath)	SEED	Sow in a lime-free peat-based compost in slight heat.	March/April	Dwarf berry-bearing evergreen shrubs for lime-free soils in sun or slight shade. Can also be divided or layered March or April.
	HALF-RIPE CUTTINGS	Root in a cold frame or propagator.	July/August	
Phellodendron (Cork Tree)	SEED	Sow in a temperature 16–18°C.	March/April	Small to medium-sized deciduous trees grown for their ornamental foliage. Grow in a sheltered position out of frost-pockets.
	HALF-RIPE CUTTINGS	Root in a propagator in slight heat.	July	
Philadelphus (Mock Orange)	SEED	Sow in slight heat.	March/April	Small to medium-sized deciduous flowering shrubs for sun or slight shade. A few species are slightly tender.
	SOFTWOOD CUTTINGS	Root in a propagator at 18°C.	May/June	
	HALF-RIPE CUTTINGS	Root in a cold frame or propagator.	July/August	
Phillyrea (Jasmine Box)	HALF-RIPE CUTTINGS	Root in a sandy compost in a cold frame or propagator with slight heat.	August	Small to medium-sized evergreen flowering shrubs or small trees for any situation. Can also be layered in summer.

PLANT NAME	METHOD OF PROPAGATION	TREATMENT REQUIRED	TIME OF YEAR PROPAGATED	CULTURAL NOTES
Phlomis (P. fruticosa = Jerusalem Sage; mild areas only)	SEED	Sow in a temperature 18–21°C.	March/April	Small evergreen flowering shrubs and herbaceous perennials. The latter can be divided spring/autumn.
	HALF-RIPE CUTTINGS	Root in a cold frame or propagator.	July/August	
Photinia (P. serrulata = Chinese Hawthorn)	SEED	Stratify overwinter then sow in a lime-free compost in a cold frame March/April.	As soon as ripe	Medium to large hardy deciduous and tender evergreen shrubs and small trees grown for their foliage and fruits. Plant in lime-free soil in sheltered positions.
	HALF-RIPE CUTTINGS	Root in a propagator at 18°C or in a cold frame in autumn.	July/August	
Phygelius (P. capensis = Cape Figwort/Fuchsia)	SEED	Sow in a temperature of 16–18°C.	March/April	Small slightly tender evergreen flowering shrubs becoming herbaceous in cold areas. Plant in sheltered positions.
	HALF-RIPE CUTTINGS	Root in a warm propagator.	June to August	
	DIVISION	Divide where possible and replant outdoors.	March/April	
Phyllodoce (Mountain Heath)	SEED	Sow in peat/sand in a shady cold frame.	March/April	Dwarf evergreen flowering shrubs for lime-free peaty soils moist but well-drained. Will grow in sun or slight shade.
	HALF-RIPE CUTTINGS	Root in a cold frame or propagator.	July/August	
Physocarpus (P. opulifolius = Nine Bark; syn. Neillia)	SEED	Sow in a seedbed outdoors or a cold frame.	March/April	Small to medium-sized deciduous flowering shrub with ornamental bark. Overwinter cuttings in the cold frame.
	HALF-RIPE CUTTINGS	Root in a cold frame.	June/July	
	HARDWOOD CUTTINGS	Root as above.	November/December	
Picea (Spruce; P. abies = Norway Spruce/Christmas Tree)	SEED	Sow in a sandy compost outdoors or cold frame.	When ripe or March/April	Evergreen conifers from dwarf prostrate species to tall trees. Varieties can also be veneer grafted in March under glass onto P. abies.
	HALF-RIPE CUTTINGS	Root in a cold frame.	September/October	
Pieris (syn. Andromeda)	SEED	Sow in sand/peat in a cold frame.	When ripe or March/April	Small to medium evergreen flowering shrubs with colourful young foliage (P. formosa forestii). Grow in sheltered spots in lime-free soils sun or slight shade.
	HALF-RIPE CUTTINGS	Root short side-branches in a cold frame. Can also be layered outdoors in spring and summer.	August/September	
Pileostegia	HALF-RIPE CUTTINGS	Root in a cold frame or propagator with slight heat.	July/August	Climbing evergreen flowering shrubs with aerial roots for cool shady positions against a wall etc.
	SERPENTINE CUTTINGS	Cut tongues and peg down as appropriate.	Spring and summer	
Pinus (Pine)	SEED	Stratify outdoors then sow outdoors or in a cold frame March/April. Protect seedlings from frost and sun.	As soon as ripe	Those cultivated here are mostly hardy coniferous trees with species from around a metre to 45 m in height. Most prefer lime-free soils.
	VENEER GRAFTING	Graft varieties onto type species in heat.	March	
Piptanthus (P. laburnifolius = Nepal Laburnum)	SEED	Sow in a sandy compost in slight heat.	March/April	Medium-sized evergreen and/or deciduous flowering shrubs best against a wall in mild areas only. P. concolor is probably the hardiest.
	HALF-RIPE HEEL CUTTINGS	Root in a cold frame and overwinter there.	August/September	
Pittosporum (Parchment Bark)	SEED	Sow in a temperature 18–21°C.	March/April	Medium to large evergreen flowering shrubs and small trees best in a frost-free greenhouse in colder areas.
	HALF-RIPE CUTTINGS	Root in a propagator at 18°C.	July/August	

PLANT NAME	METHOD OF PROPAGATION	TREATMENT REQUIRED	TIME OF YEAR PROPAGATED	CULTURAL NOTES
Platanus *(Plane)*	SEED˙	Sow in a seedbed outdoors or cold frame.	When ripe or March/April	Large deciduous trees with ornamental leaves and bark for deep moist soils. Can be pruned to shape in winter.
	HARDWOOD HEEL CUTTINGS	Root 20–30 cm cuttings in a cold frame. Can also be air layered in summer.	October to December	
Podocarpus *(P. andinus =* *Plum-fruited Yew)*	SEED	Sow in a temperature of 18–21°C.	March/April	Evergreen trees and shrubs from dwarfs to large species. *P. andinus* is the hardiest but all are tender to some degree.
	HALF-RIPE CUTTINGS	Root in a propagator or frost-free frame.	July/August	
Polygonum *(P. baldschuanicum =* *Russian Vine)*	HARDWOOD CUTTINGS	Root in a cold frame or frost-free greenhouse.	November/December	Vigorous deciduous ornamental flowering climbers for trellises etc.
Poncirus *(Japanese Bitter Orange)*	SEED	Place out to stratify over winter then sow in a cold frame in March/April.	As soon as ripe	Medium-sized spiny deciduous flowering shrubs related to *Citrus*. Can be trimmed to form hedges.
	HALF-RIPE CUTTINGS	Root in a propagator at 18°C.	August/September	
Populus *(Poplar)*	HARDWOOD CUTTINGS	Root in a sheltered spot outdoors.	November/December	Large fast-growing deciduous trees for moist soils and ideal for forming screens, especially *P. deltoides* and *P. nigra italica*.
	DIVISION (suckers)	Separate and replant. Species can also be raised from seed sown outdoors when ripe.	March/April	
Potentilla *(Shrubby Cinquefoil)*	SEED	Sow in a temperature of 16–18°C.	March/April	Shrubby types are small spreading deciduous plants with many varieties all flowering over a long period.
	HALF-RIPE CUTTINGS	Root in a cold frame.	September/October	
Prunus *(For fruit and ornamental grafting see Chapter 13)*	SEED	Stratify over winter then sow outdoors or in a cold frame March/April.	As soon as ripe	Very varied genus of deciduous flowering and/or fruiting trees and a few evergreens (Cherry Laurels), the latter best in sheltered positions and lime-free soils.
P. lusitanica and *P. laurocerasus*	HALF-RIPE CUTTINGS	Root in a cold frame.	September/October	
P. pumila and *P. glandulosa*	SOFTWOOD CUTTINGS	Root in a propagator at 18°C.	May/June	
Pseudolarix *(Golden Larch)*	SEED	Sow in a seedbed outdoors or in a cold frame.	When ripe or March/April	Tall deciduous coniferous trees for lime-free soils. Good autumn foliage.
Pseudotsuga *(Douglas Fir)*	SEED	Stratify outdoors over winter then sow in a seedbed outdoors or in a cold frame.	As soon as ripe	Tall evergreen coniferous tree for lime-free, moist but well-drained soils.
Ptelea *(Hop Tree)*	SEED	Stratify outdoors over winter then sow in a cold frame or outdoor seedbed.	As soon as ripe	Small deciduous flowering trees for sun or slight shade. Can also be layered in spring and summer.
Pterocarya *(Wingnut)*	SEED	Stratify outdoors over winter then sow in a cold frame or seedbed outdoors March/April.	As soon as ripe	Medium to large deciduous trees grown for their foliage and catkin-like flowers. Grow well in deep moist soils, especially near water.
	DIVISION (suckers)	Separate where formed and replant. Can also be layered in summer.	March	
Punica *(P. granatum =* *Pomegranate)*	SEED	Sow in a temperature of 16–18°C.	January to April	Small deciduous flowering tree with edible fruits best grown against a warm wall in mild areas.
	HALF-RIPE CUTTINGS	Root in a propagator at 18°C.	June/July	

PLANT NAME	METHOD OF PROPAGATION	TREATMENT REQUIRED	TIME OF YEAR PROPAGATED	CULTURAL NOTES
Pyracantha *(P. coccinea = Fire Thorn)*	SEED	Stratify outdoors over winter then sow in a cold frame.	As soon as ripe	Medium-sized evergreen flowering and berry-bearing thorny shrubs; a few species are slightly tender.
	HALF-RIPE CUTTINGS	Root in a cold frame or propagator.	August	
Pyrus *(for grafting fruit species etc, see Chapter 13)*	SEED	Stratify outdoors over winter then sow in a cold frame or outdoor seedbed in March/April.	As soon as ripe	Small to medium-sized deciduous flowering and fruit-bearing trees.
Quercus *(Oak)*	SEED (acorns)	Sow outdoors where they will grow or singly in pots in a cold frame.	As soon as ripe	Varied genus of evergreen and deciduous trees and a few shrubs. Deciduous varieties are spiced veneer grafted onto their species in March under glass.
	HALF-RIPE CUTTINGS (evergreens)	Root in a propagator at 18°C.	July/August	
Raphiolepis *(R. indica = Indian Hawthorn)*	HALF-RIPE CUTTINGS	Root in a slightly heated propagator in John Innes seed compost.	July/August	Small slow-growing evergreen flowering shrubs for milder areas, *R. umbellata* being the hardiest.
Rhamnus *(Buckthorn)*	SEED	Stratify outdoors over winter then sow outdoors or in a cold frame in March/April.	As soon as ripe	Garden types are medium sized berry-bearing evergreen shrubs or small trees for sun or slight shade. Good shrubs for coastal areas.
	HALF-RIPE CUTTINGS	Root in a cold frame or propagator with slight heat.	July/August	
Rhododendron *(includes Azaleas)*	SEED (surface sow)	Sow on sand/peat and cover lightly with sand. Germinate in a propagator at 13–16°C.	March/April	Dwarf to large evergreen and deciduous flowering shrubs and trees for lime-free soils with many species and varieties some needing greenhouse protection. Specialist books should be consulted for individual cultural requirements. All outdoor types can be layered outdoors in summer.
	HALF-RIPE CUTTINGS (shrubby types)	Root in sandy peat in a propagator or under mist at 16–18°C. Smaller leaf types root the easiest.	July to October	
	SADDLE GRAFTING	Graft hybrid evergreens under glass onto *R. ponticum*. Deciduous azaleas onto *R. luteum*.	March	
Rhodotypos *(White Kerria)*	HALF-RIPE CUTTINGS	Root in a propagator at 16–18°C.	June/July	Small deciduous flowering shrub with black berries in winter. Plant in sun or slight shade.
	HARDWOOD CUTTINGS	Root in a cold frame.	October/November	
Rhus *(Sumach; some species are poisonous to allergic people, e.g. R. radicans = Poison Ivy)*	SEED	Sow in slight heat.	March/April	Deciduous flowering trees shrubs and climbers mostly grown for their autumn foliage and fruits. Some species need a greenhouse.
	ROOT CUTTINGS	Root 5–8 cm cuttings in 8 cm pots in a cold frame.	December/January	
	DIVISION (suckers)	Separate and replant according to species. Can also be layered in autumn.	In autumn or early spring	
Robinia *(False Acacia/Locust)*	SEED (soak 12 hours in warm water before sowing)	Sow in a cold frame or in slight heat in the greenhouse.	March/April	Deciduous flowering trees and shrubs with a range of sizes all with attractive foliage. Plant in hot, dry sheltered positions. Varieties are whip and tongue grafted onto *R. pseudoacacia*.
	ROOT CUTTINGS	Root in a cold frame.	December/January	
	DIVISION (suckers)	Separate and replant in a nursery bed to grow on.	March/April	
Rubus *(Ornamental Bramble)* *R. odoratus, spectabilis and parviflorus*	LAYERING (tip and ordinary)	Peg down appropriate shoots outdoors.	Summer	Many flowering and/or ornamental species and varieties of evergreen or deciduous shrubs and climbers as well as the familiar blackberry. Grow in sun or shade.
	DIVISION	Divide and replant according to the species.	March/April	
R. cockburnianus	ROOT CUTTINGS	Grow on in a cold frame.	December/January	

PLANT NAME	METHOD OF PROPAGATION	TREATMENT REQUIRED	TIME OF YEAR PROPAGATED	CULTURAL NOTES
Ruscus (R. aculeatus = Butcher's Broom)	DIVISION	Replant in sun or shade which can be quite dense. Can also be raised from seed.	March/April	Small evergreen ornamental berry-bearing shrubs with male and female plants.
Salix (Willow/Sallow)	HARDWOOD CUTTINGS	Root in a sheltered spot outdoors. Can also be raised from seed and layers.	November/December	Deciduous trees and shrubs with a wide range of sizes. Grown for their foliage and catkins.
Santolina (S. chamaecyparissus = Lavender Cotton)	HALF-RIPE CUTTINGS	Root in a cold frame.	September/October	Very small evergreen aromatic flowering/foliage shrubs for light sandy soils. Cut back annually.
	HARDWOOD CUTTINGS	Root in a cold frame or sheltered spot outdoors.	November/December	
Sarcococca (Sweet Box)	SEED	Sow outdoors or in a cold frame.	March/April	Small, often low-growing evergreen shrubs with fragrant flowers. Plant in shady positions especially under trees.
	HALF-RIPE CUTTINGS	Root in a cold frame.	October	
	DIVISION	Replant according to species.	March/April	
Sassafras (S. albidum = Sassafras Tree)	SEED	Sow in pots and place out to stratify then move to slight heat in February/March.	December/January	Large deciduous trees grown for their aromatic leaves. Plant in sheltered positions.
	ROOT CUTTINGS	Root in a greenhouse or cold frame.	December/January	
Sciadopitys (Umbrella Pine)	SEED	Sow in slight heat or a cold frame.	March/April	Large slow-growing evergreen conifer for sheltered positions in lime-free soils. Sun or slight shade.
Sequoia (Californian Redwood)	SEED	Sow in a sandy compost in a cold frame.	March/April	Very large evergreen coniferous trees for sheltered positions in deep soils. Forms can be grafted onto the species.
	HALF-RIPE CUTTINGS	Root in a propagator at 18°C.	September/October	
Sequoiadendron (Wellingtonia)	SEED	Place out to stratify then sow in a cold frame or in slight heat March/April.	January/February	Tall evergreen coniferous tree with reddish-brown spongy bark.
Skimmia	SEED	Sow in a cold frame.	When ripe or March/April	Small evergreen shrub grown for flowers/fruit and foliage. Some species have male and female plants. Grow in lime-free soils in sun or slight shade.
	HALF-RIPE CUTTINGS	Root in a cold frame. Can also be layered outdoors in spring or summer.	September/October	
Sophora (S. japonica = Japanese Pagoda Tree)	SEED (soak 24 hours in warm water before sowing)	Sow in a sandy compost in a temperature of 18–21°C.	March/April	Small to medium-sized deciduous and evergreen flowering trees and shrubs, the evergreens being slightly tender and best in mild areas.
Sorbaria (False Spiraea)	DIVISION (suckers)	Separate and replant in rich moist soils.	Spring or autumn	Small to medium-sized deciduous flowering shrubs. They can also be raised from seed when available.
	HARDWOOD CUTTINGS	Root in a cold frame.	November/December	
	ROOT CUTTINGS	Root and grow on in a cold frame.	December/January	
Sorbus (S. aucuparia = Rowan, Mountain Ash; S. aria = Whitebeam)	SEED	Sow outdoors or stratify and move to a cold frame March/April.	As soon as ripe	Small ornamental deciduous trees grown for their flowers and fruits. Whitebeams grow well on chalk soils.
	BUDDING	Bud varieties onto their species.	July	
Spartium (S. junceum = Spanish Broom)	SEED	Sow in pans in a cold frame or greenhouse. Grow on in pots until planted out.	March/April	Deciduous flowering shrubs 1.8–3 m in height. Ideal for dry sites in coastal areas.

PLANT NAME	METHOD OF PROPAGATION	TREATMENT REQUIRED	TIME OF YEAR PROPAGATED	CULTURAL NOTES
Spiraea (S. arguta = Bridal Wreath/Foam in May)	SOFTWOOD CUTTINGS	Root in a cold frame or propagator.	June/July	Dwarf to small deciduous flowering shrubs with numerous species some of which can also be divided in spring.
	HARDWOOD CUTTINGS	Root in a cold frame or sheltered spot outdoors.	November/December	
Stachyurus	HALF-RIPE CUTTINGS	Root in a propagator at 18°C. Can also be layered outdoors in summer.	July/August	Small to medium-sized deciduous flowering shrubs for sun or slight shade.
Staphylea (Bladder Nut)	SEED	Sow in slight heat. Seed may benefit from warm/cool stratification.	March/April	Small to medium-sized deciduous shrubs grown for their ornamental foliage, flowers and fruit. Grow in sun or slight shade. Can also be layered in summer.
	HALF-RIPE CUTTINGS	Root in a warm propagator.	July/August	
	DIVISION (suckers)	Separate and replant in moist soils.	Autumn or early spring	
Stephanandra	HALF-RIPE CUTTINGS	Root in a cold frame or propagator.	July	Small deciduous shrubs grown for their ornamental foliage and stems. Grow in sun or slight shade especially near water.
	HARDWOOD CUTTINGS	Root in a cold frame or sheltered spot outdoors.	November	
	DIVISION	Replant in moist soils.	March/April	
Stewartia	SEED	Sow in a lime-free compost in slight heat.	March/April	Medium to large deciduous shrubs grown for flowers, autumn foliage and ornamental bark. Sheltered positions in slight shade and lime-free soils.
	SOFTWOOD CUTTINGS	Root in a propagator at 18°C.	June	
Stranvaesia	SEED	Stratify outdoors over winter then sow outdoors or in a cold frame March/April.	As soon as ripe	Medium-sized evergreen shrubs or small trees grown for their foliage, flowers and berries. Grow in sun or slight shade.
	LAYERING	Peg down appropriate branches.	Spring and summer	
Styrax (Snowbell)	SEED	Sow in a peaty soil in slight heat. Warm/cool stratification may be of benefit.	March/April	Medium to tall deciduous flowering shrubs, some rather tender and best grown in mild areas. All are better in lime-free soils.
	HALF-RIPE CUTTINGS	Root in a propagator at 18°C.	June/July	
	AIR LAYERING	Layer as described on page 29.	Spring and summer	
Sycopsis	SEED	Sow in slight heat.	March/April	Small evergreen winter flowering trees for sun or slight shade.
	HALF-RIPE CUTTINGS	Root in a warm propagator.	July/August	
Symphoricarpos (Snowberry)	SEED	Stratify outdoors then sow in a cold frame March/April.	As soon as ripe	Small deciduous flowering and berry-bearing shrubs for sun or slight shade. Suckers sometimes form and can be detached to replant.
	HARDWOOD CUTTINGS	Root in a cold frame or sheltered spot outdoors.	November/December	
Syringa (Lilac)	SEED	Sow in pots placed out to stratify and move to a cold frame in March/April.	December/January	Deciduous flowering shrubs in a wide range of heights and growing well on chalk. Varieties can be grafted onto *Ligustrum* but are better on own roots.
	HALF-RIPE CUTTINGS	Root in a propagator at 18°C. Can also be layered in spring and summer.	August/September	
Tamarix (Tamarisk)	HARDWOOD CUTTINGS	Root in a cold frame or sheltered spot outdoors.	October to December	Small to medium-sized deciduous flowering shrubs growing best in coastal areas on neutral soils.

PLANT NAME	METHOD OF PROPAGATION	TREATMENT REQUIRED	TIME OF YEAR PROPAGATED	CULTURAL NOTES
Taxodium (Swamp Cypress)	SEED	Sow in pots placed out to stratify and move to a cold frame March/April.	December/January	Medium to large deciduous coniferous trees especially good for moist soils at the waterside.
	HARDWOOD CUTTINGS	Root in a propagator with slight heat.	November/December	
Taxus (Yew; leaves poisonous to cattle)	SEED	Stratify outdoors then sow outdoors or in a cold frame in March/April.	As soon as ripe	Evergreen trees or tall shrubs for well-drained positions in sun or shade. T. baccata is used in hedging and topiary. Golden forms are veneer grafted onto T. baccata under glass.
	HALF-RIPE CUTTINGS	Root in a cold frame or propagator.	August to October	
Telopea (T. truncata = Tasmanian Waratah)	SOFTWOOD CUTTINGS	Root in a propagator at 18°C and grow on under glass.	June/July	Evergreen flowering shrubs and trees, T. truncata being the hardiest. Grow in lime-free soils in mild areas.
	LAYERING	Cut tongue and layer into pots where possible.	Autumn and winter	
Thuja (Arbor-vitae and **Thujopsis**; Hiba Arbor-vitae)	SEED	Sow in pots, place out to stratify and move to a cold frame March/April.	December/January	Evergreen coniferous trees and shrubs with many forms and varieties. T. occidentalis and T. plicata are used as hedge plants. Plant in sun or slight shade.
	HALF-RIPE CUTTINGS	Root in a cold frame or propagator.	September/October	
Tilia (Lime/Linden)	SEED	Sow outdoors or in a cold frame. May benefit from warm/cool stratification.	As soon as ripe	Medium to large deciduous trees useful for screens etc as it is easily pruned and shaped.
	AIR LAYERING	Layer as described on page 29.	April to June	
Torreya (Stinking Yews)	SEED	Sow in a sandy compost in a temperature of 18–21°C.	March/April	Large evergreen trees or bushes for sheltered positions in sun or shade in mild areas only.
	HALF-RIPE CUTTINGS	Root short cuttings in a cold frame or propagator.	August/September	
Trachelospermum	HALF-RIPE CUTTINGS	Root in a propagator at 18°C.	August	Evergreen twining climbers with jasmine-like flowers. Grow on a warm wall or T. jasminoides in a cool greenhouse in cold areas.
	SERPENTINE LAYERING	Cut tongues and peg into pots as appropriate.	Spring and summer	
Trachycarpus (Chusan Palm)	SEED	Sow in a temperature of 18–21°C.	March/April	T. fortunei is a palm with fan-shaped leaves hardy in most parts, only needing shelter from cold winds.
Trochodendron	SEED	Sow in slight heat.	Autumn or spring	Small but spreading ornamental evergreen trees.
	AIR LAYERING	Layer as described on page 29.	Spring and summer	
Tsuga (Hemlock)	SEED	Sow in pots, place out to stratify and move to a cold frame in March/April.	When ripe or December/January	Tall coniferous evergreen trees for sun or shade in deep moist soils. T. heterophylla is very hardy but some are best suited to mild areas.
	HALF-RIPE CUTTINGS	Root in a cold frame. Varieties can also be grafted under glass in spring.	July to September	
Ulex (Gorse/Furze)	SEED	Sow where they are to grow or in pots under glass.	March/April	Dwarf and small very spiny evergreen flowering shrubs best on poor dry soils but not over chalk. Avoid root disturbance when planting out.
	HALF-RIPE CUTTINGS (for double forms)	Root in a cold frame or propagator. Grow on in pots until planting out.	August	

PLANT NAME	METHOD OF PROPAGATION	TREATMENT REQUIRED	TIME OF YEAR PROPAGATED	CULTURAL NOTES
Ulmus *(Elm)*	SEED	Sow in a seedbed outdoors.	As soon as seed is ripe	Mostly very large deciduous trees best in large open areas rather than gardens. Forms can also be budded or grafted onto *U. glabra* (Wych Elm).
	SOFTWOOD CUTTINGS	Root in a propagator at 18°C.	June	
	DIVISION (suckers)	Separate where they form and replant. Can also be layered.	Autumn and winter	
Umbellularia *(Californian Laurel; foliage can cause skin irritation to allergic people)*	SEED	Sow in slight heat.	March/April	Large aromatic evergreen shrub or small tree grown for its flowers and large ornamental fruits. Hardy in most areas.
	HALF-RIPE CUTTINGS	Root in a cold frame or propagator. Can also be layered outdoors in spring or summer.	July/August	
Vaccinium *(Blueberry/Billberry/ Cranberry)*	SEED	Sow in moist sandy peat in slight heat.	March/April	Dwarf to medium-sized flowering deciduous and evergreen shrubs for moist lime-free soils in slight shade. Bear edible berries.
	HALF-RIPE CUTTINGS	Root 10 cm cuttings in a cold frame or propagator.	August	
Viburnum	SEED	Sow in a cold frame or greenhouse.	March/April	Mostly small to medium-sized evergreen and deciduous flowering shrubs some grown for their ornamental fruits or autumn foliage. Grow in sun or slight shade.
	HALF-RIPE CUTTINGS	Root in a cold frame or propagator with slight heat.	July to September	
	HARDWOOD CUTTINGS	Root in a cold frame. Can also be layered in spring and summer.	November/December	
Vinca *(Periwinkle)*	HALF-RIPE CUTTINGS	Root in a cold frame.	August to October	Trailing evergreen and deciduous flowering shrubs making good ground cover especially in sun. *V. difformis* is slightly tender.
	DIVISION	Replant according to species. Spread indefinite.	March to May	
Viscum *(V. album = Mistletoe)* *(Other hosts: Poplars, Limes, Maples and Mountain Ash)*	SEED	Press ripe berry against underside of a young branch of hawthorn (*Crataegus*) or apple (*Malus*) for preference.	May	Evergreen parasitic berry-bearing shrubs with male and female plants both being needed in close proximity for berry production.
Vitex *(V. agnus-castus = Chaste Tree)*	SEED	Stratify in cool conditions then sow at 18–21°C in March/April.	December/January	Deciduous aromatic flowering shrubs up to 3 m in height. Grow against a warm wall in milder areas.
	HALF-RIPE CUTTINGS	Root in a propagator at 16–18°C.	July/August	
Weigela	HALF-RIPE CUTTINGS	Root in a cold frame or propagator.	June to August	Mostly small deciduous flowering shrubs for sun or slight shade. Species can also be raised from seed.
	HARDWOOD CUTTINGS	Root in a cold frame.	November/December	
Wisteria	SEED (soak 12 hours in warm water before sowing)	Sow outdoors or in a temperature of 16–18°C.	March/April	Deciduous flowering climbers which can be trained as shrubs and standards (*W. sinensis*) or against walls, over trees etc.
	HARDWOOD CUTTINGS	Root in a frost-free greenhouse.	November/December	
	SERPENTINE LAYERING	Layer into pots as appropriate.	Spring and summer	
Xanthoceras	SEED	Sow outdoors or in a cold frame.	When ripe or March/April	Small deciduous flowering tree best on neutral or acid soils in a shady position.
	ROOT CUTTINGS	Root in a frost-free greenhouse in a sandy compost.	December/January	

PLANT NAME	METHOD OF PROPAGATION	TREATMENT REQUIRED	TIME OF YEAR PROPAGATED	CULTURAL NOTES
Yucca	SEED (hand pollinate for own seed)	Sow in a temperature of 13–16°C. or more for tender species.	March/April	Unusual small shrubs or trees with ornamental evergreen foliage and exotic flowers. Amongst hardy species are *Y. filamentosa, flaccida* and *recurvifolia*. Many others need frost protection.
	DIVISION (suckers)	Separate where formed and replant indoors or out.	March/April	
	ROOT CUTTINGS	Root 5–8 cm cuttings in a propagator at 16–18°C.	March/April	
Zanthoxylum (Prickly Ash)	SEED	Sow in slight heat. May benefit from warm/cool stratification.	March/April	Medium-sized ornamental deciduous shrubs or small trees for sun or slight shade. Most are spiny and aromatic.
	ROOT CUTTINGS	Root in a greenhouse or cold frame.	December/January	
Zelkova	SEED	Sow outdoors or in a cold frame.	When ripe or March/April	Large deciduous elm-like trees for deep moist soils especially near water.
	AIR LAYERING	Layer as described on page 29.	Spring and summer	
Zenobia	SEED	Sow in a cold frame.	March/April	Small deciduous or semi-evergreen flowering shrub for lime-free soils and slight shade.
	HALF-RIPE CUTTINGS	Root in a sandy compost in a propagator at 18°C. Can also be layered in summer.	July/August	

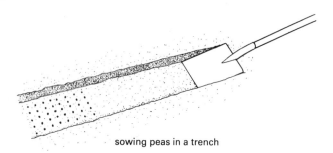

sowing peas in a trench

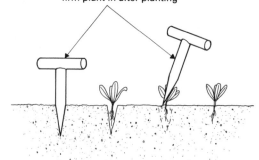

use dibber to make holes and firm plant in after planting

Planting out cabbages

Crop Rotation

Plot A Manure or compost	Plot B Fertilizer only		Plot C Fertilizer only	Plot D Fertilizer, manure or compost
Brassicas (cabbage family) Brussels sprouts Cabbages Cauliflowers Potatoes (in small gardens)	Peas, beans, salad crops etc Beans Celery Leeks Lettuces Marrows Onions	Peas Shallots Spinach Sweetcorn Tomatoes	Root crops Beetroots Carrots Parsnips Radishes Swedes	Potatoes (in large gardens)

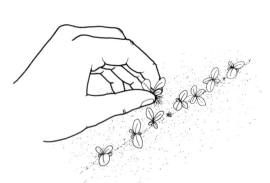

thinning lettuce seedlings

press seed into compost

sow two seeds in a fibre pot and thin to one plant later if necessary

plant seedling out complete with pot which will rot away

Cucumbers

19 Vegetables

Most vegetables are raised from seed, many of which can be self-saved by the amateur, and most are not difficult to grow if their main requirements of full sunlight and well cultivated humus-rich soil are provided. Most do, however, take up a great deal of space, and as it is also advisable to rotate most crops to fresh ground on a 3–4 year cycle (see opposite) a large area needs to be set aside for them.

For smaller gardens it is more economic to concentrate on salad plants as these take less space than other vegetables and also less time to mature, making a number of crops per year possible. Many also tolerate and grow well in slight shade provided that the soil is not too dry. With these as with other vegetables the use of cloches, frames and a greenhouse will extend and enhance the growing/harvesting seasons of many plants being used either to raise plants earlier in spring or to protect maturing crops in autumn and winter. Some can, of course, spend their entire lives under glass.

Correct cultivation will lessen the likelihood of pest and disease, but where these do strike quick action with appropriate sprays etc is essential for a satisfactory crop.

For winter storage the modern freezer has solved many problems, but unless you insist on runner beans in January a freezer is by no means essential to a varied year-round vegetable supply. Much can be achieved by sowing winter/spring maturing varieties and storing root crops in sand boxes indoors or in soil 'clamps' outside. Again, cloches and polythene tunnels will keep the worst of winter from growing crops, excessive wet and wind being a greater enemy than frost in many cases.

Most brassicas (cabbages etc) are raised in nursery beds, sowing thinly in drills 15 cm or so apart, and transplanted to their final positions a couple of months later. Plants raised under glass should be pricked out/ potted on as for other seedlings, and always remember to harden off before planting out.

The crop rotation plan opposite can be adapted to either 3 years or 4, depending on the size of the vegetable garden. In the second year Plot A should be planted with the vegetables listed here for Plot B, Plot B with those for Plot C, Plot C with those for Plot A, and so on for subsequent years. In large gardens a 4-year rotation should be adopted, incorporating Plot D into the plan.

In the cultural notes following I have indicated the parts of the vegetable used and how, the months mentioned being the seasons when the plant can be harvested. Sowing/planting distances are given as average; some giant or dwarf varieties may need more or less space as the case may be. They may not give the heaviest crop for a given area, but take into account kitchen needs and ease of cultivation.

PLANT NAME	METHOD OF PROPAGATION	TREATMENT REQUIRED	TIME OF YEAR PROPAGATED	CULTURAL NOTES
Artichoke, Globe	DIVISION (suckers)	Detach with heel when 23 cm long. Replant 10 cm deep, 90 cm apart.	April	Flowerheads used cooked. Cut just before they open July to October. Plant seedlings into a cropping bed in spring.
	SEED (sow 1.3 cm deep)	Sow in rows 30 cm apart, and thin to 15 cm apart in the row.	Late May/early June	
Artichoke, Jerusalem	TUBERS	Large tubers can be divided. Plant 15 cm deep, 38 cm apart, in rows 75 cm apart.	March/April	Tubers used cooked. Tall plants also useful as a windbreak. Herbaceous perennial. Lift as required November to March.
Asparagus	CROWNS	Plant in a trench 10 cm deep, 15 cm wide, planting 45–60 cm apart both ways.	March/April	Young shoots are used cooked, new beds not being cropped for their first 3 years. Remove any flowers as they form. In season April to June and can be forced. Plant seedlings out in spring.
	SEED (soak 12 hours in warm water before sowing)	Sow 2.5 cm deep, 2.5 cm apart, in drills outdoors 15 cm apart and thin to 8 cm in the row.	Early April	
Asparagus Pea	SEED (sow 2.5 cm deep)	Sow in drills 50–60 cm apart and thin to 45 cm in the row. Can also be sown early under glass to plant out mid-May.	April	Pods are cooked whole when young. Grow in warm mild areas only. In season June to September.
Aubergine	SEED (just cover)	Sow in a temperature of 18–21°C. Prick out singly into small peat pots. Plant out 45 cm apart.	January to March	Best grown under glass in growing-bags or harden off and plant in warm spot outdoors early June. Fruits used raw or cooked July to October.
Beetroot	SEED (sow 2.5 cm deep)	Sow thinly in rows 30 cm apart, and thin to 8 cm, then 15 cm, using second thinnings for the table.	April to July	Roots used cooked and served hot or cold. July onwards being stored overwinter in sand or outdoor 'clamps'.
Borecole/Kale	SEED (sow 1.3 cm deep)	Sow in drills in a nursery bed. Plant out June/July 45 cm apart, 60 cm between rows.	April/May	Central leaves used a few at a time cooked. In season December to April.
Broad Bean	SEED (sow 5–6 cm deep)	Sow in double drills 20 cm apart with seed planted alternately at the same distance. Space double drills 75 cm apart.	Mid-February to mid-April or November	Seed used cooked as soon as large enough. May to July. November sowings can be grown on under cloches.
Broccoli, Purple-sprouting	SEED (sow 1.3 cm deep)	Sow in a nursery bed and plant out 45 cm apart in rows 75 cm apart in June.	April/May	Flowerbud sideshoots used cooked when 15 cm long. March to June.
Broccoli, Spring or Heading/Winter Cauliflower	SEED (sow 1.3 cm deep)	Sow and plant as above.	April/May	Use heads as cauliflower. Choose hardy varieties for cold areas. February to May.
Brussels Sprouts	SEED (sow 1.3 cm deep)	Sow in a nursery bed. Plant out in May/June 60 cm apart each way.	February to April	Can be in season from August through to April. Tops and flower buds are also used cooked.
Cabbage, Spring	SEED (sow 1.3 cm deep)	Sow all in seedbeds. Transplant September/October. Transplant in April.	Sow: August	Mostly used cooked but some are used raw in winter salads etc. Seasons in use. **Spring**, April to June. **Late summer**, September to December. **Winter**, December/January.
Late Summer			March	
Winter		Transplant June/July. Plant out 45 cm apart in rows 60 cm apart.	April	

PLANT NAME	METHOD OF PROPAGATION	TREATMENT REQUIRED	TIME OF YEAR PROPAGATED	CULTURAL NOTES
Cabbage, Red	SEED (sow 1.3 cm deep)	Sow in a nursery bed. Transplant in spring 60 cm apart.	September	Used for pickling or raw in salads. September to March.
Cabbage, Savoy	SEED (sow 2 cm deep)	Sow in a nursery bed. Transplant 60 cm apart each way from late June to August.	April/May	Very hardy cabbages in season from October to March.
Calabrese	SEED (sow 1.3 cm deep)	Sow in a nursery bed. Transplant 60 cm apart each way in June/July.	April/May	Flowerheads and stems used while in tight bud cooked. Can be started early under glass, August to November.
Cardoon	SEED	Sow in a cold frame or greenhouse. Plant out early May 50 cm apart in trenches as for celery.	In spring	Blanched stems used cooked or raw in winter. Tie leaves together in autumn and earth up. Protect from frost in cold areas.
Carrots	SEED (sow 0.6 cm deep)	Sow under cloches: in drills 23 cm apart. Cold frames: or outdoors in drills 25–30 cm apart. Thin to 10 cm in the row.	March or September February to April	Roots used cooked or raw. Do not grow on freshly manured ground. Can be used throughout the year using cloches and winter storage.
Cauliflower	SEED (sow 1.3 cm deep)	Sow in a seedbed. Transplant 60 cm apart each way in May. Can be raised early under glass.	April	Flower heads (curds) are used cooked when well formed. Can be sown in a cold frame September to plant out in April. June to October.
Celeriac (Turnip Rooted Celery)	SEED	Sow in a peat-based compost at 16°C. Plant out 30 cm apart in rows 45 cm apart in May.	March	Use swollen roots cooked or raw in salads. Lift and store in sand in autumn. October to March.
Celery	SEED (wash in cold clean water before sowing)	Sow in a temperature of 13–16°C. Plant out in June 25–30 cm apart, self-blanching types in rows, autumn/winter varieties in prepared trenches 15 cm deep.	February to April	Use stems cooked or raw in salads. Seed germinates well under mist. Protect winter types from frost where possible. August onwards.
Celery Mustard (Mustard Spinach/Pak Choi)	SEED (sow 1.3 cm deep)	Sow in drills 38 cm apart and thin to 38–45 cm in the row.	April onwards for succession	Leaves used cooked though hearts can be eaten raw in salads and some are grown for their flower stems. Some type can be had all year round.
Celtuce	SEED (sow 1.3 cm deep)	Sow in drills 38 cm apart and thin to 23 cm in the row.	April to July (fortnightly for succession)	Leaves and heart used cooked or raw in salad. June to October.
Chinese Cabbage	SEED (sow 1.3 cm deep)	Sow in rows 30 cm apart and thin to 23 cm in the rows.	May to July	Use leaves cooked like cabbage or raw like lettuce. August/September.
Courgettes (Zucchini Marrows)	SEED (sow 1.3 cm deep)	Sow singly in small pots at 16°C. Plant out 30 cm apart in rows 60 cm apart. Can also be sown outdoors.	May	Fruits used cooked when 5–15 cm long depending on the variety. Pollinate artificially for best results. June to September.
Cucumber	SEED (sow 1.3 cm deep on edge)	Sow 2 seeds together in 8 cm pots at 16–18°C. Thin to strongest plant later. Plant out 60 cm apart. Sow ridge types as above in March and plant out in June or sow outdoors late May.	February to May	Fruits used raw in salads. Grow frame types under glass, ridge types outdoors. Remove any male flowers from frame types but leave *on* ridge. July to September.
Endive	SEED (sow 1.3 cm deep)	Sow in rows 30 cm apart and thin to 23 cm in the row. Sow every 3–4 weeks for succession.	May to September	Salad plant. Blanch by tying leaves together for a day or two before harvesting. Autumn and winter.

PLANT NAME	METHOD OF PROPAGATION	TREATMENT REQUIRED	TIME OF YEAR PROPAGATED	CULTURAL NOTES
Haricot and **French Beans**	SEED (sow 5 cm deep)	Sow in rows 45 cm apart spacing seed 8–10 cm apart and thinning to 15–20 cm in the row.	May to July	Dwarf or climbing beans. Use pods of French, seeds of haricots cooked. July to October. Can be started off early under glass.
Kohl Rabi	SEED	Sow in rows 30–38 cm apart. Thin to 23 cm in the row. Sow every 3 weeks for succession.	March to May	Swollen stems used when tennis-ball-sized cooked. Can be started off early under glass. May onwards.
Leaf Beet *(Perpetual Spinach/Spinach Beet)*	SEED (sow 2.5 cm deep)	Sow in rows 45 cm apart and thin to 20–23 cm in the row.	April and July/August	Use leaves and stems cooked. Can be harvested all year with successional sowings.
Leek	SEED (sow 1.3 cm deep)	Sow in a temperature of 13–16°C. Plant out in rows 30–46 cm apart in June/July spacing plants every 15–23 cm.	March	Blanched stem used cooked. Plant out with a dibber in holes 23 cm deep, simply dropping plants in and watering. Harvest as required. December to May.
Lettuce	SEED (sow 1.3 cm deep)	Sow all in rows 30 cm apart and thin to 23 cm in the row.		Salad crop. Choose varieties carefully to suit time of year and method of growing i.e. cloches etc. Can be cropped all year with careful cultivation.
		Summer and autumn.	March to August	
		Winter (outdoors).	August/September	
		Winter (cloche).	October/November	
Marrow *(includes Squashes and Vegetable Spaghetti)*	SEED (sow 1.3 cm deep)	Sow singly in small pots under glass or outdoors where they will grow in late May/early June. Sow/plant out bush types 90 cm apart, trailing types 1.5 m.	March/April	Fruits eaten cooked. Can be grown under cloches or in frames if wished. Hand pollinate for best results. July to December.
Mushrooms	SPAWN	Grow in large 23–30 cm deep boxes or in beds as described on page 22.	According to cropping time required	Fruiting bodies eaten cooked. Can be harvested all year with correct cultivation.
Mustard and Cress	SEED (surface sow)	Sow in shallow containers in a warm position. A wad of cotton wool or similar will do to sow on.	At any time	Use in salads when 5–8 cm high. When used together sow mustard 3 days later than cress. Commercially *Brassica napus* (Rape) is used.
Okra	SEED (sow 5 cm deep)	Sow in rows 60 cm apart and thin to 45 cm in the row. Can be started early under glass. Protect seedlings from frost.	Early May	Fruits are used cooked. Grow outdoors in mild sunny areas elsewhere in a greenhouse as for tomatoes. August to October.
Onion	SEED (sow 1.3 cm deep)	Sow in rows 30 cm apart and thin to 23 cm in the row.	March/April	Bulbs used cooked or raw. In colder areas or for exhibition use seed in best started off under glass in January at 10°C. Grow on in small pots and plant out in May. With the right varieties and winter storage can be available all year.
	SETS (bulblets)	Plant 15 cm apart in rows as above just pressing them into soil surface.	March/April	
Winter-hardy Japanese Types	SEED	Sow in rows 23 cm apart and thin to 5 cm in the row in spring.	August (only)	
Spring onions	SEED	Sow in rows 30 cm apart at 2 week intervals and pull as needed.	In spring and July/August	Use thinnings of Japanese types as spring onions.
Onion (Tree)	SETS (bulblets)	Plant 2.5 cm deep, 23 cm apart, in groups as formed or singly.	September	Bulblets form in bunches at top of the stem. Use cooked or raw.
Parsnip	SEED (sow 1.3 cm deep)	Sow in rows 38 cm apart and thin to 23 cm in the row. Always use fresh seed.	March	Root used cooked. Leave in the ground and harvest as required or lift and store. November onwards.

PLANT NAME	METHOD OF PROPAGATION	TREATMENT REQUIRED	TIME OF YEAR PROPAGATED	CULTURAL NOTES
Peas	SEED (sow 5 cm deep)	Sow dwarf types in drills, taller ones in 15 cm wide, 5–8 cm deep, trenches. Space rows from 30 cm to 90 cm apart according to variety height.	November to July sowing round seed varieties first	Seed used cooked. Tall types need the support of sticks or netting. Dwarf types can be grown under cloches. May to September.
Peppers	SEED	Sow in a temperature of 16–21°C. Prick out into small fibre pots and grow on in slight heat. Plant out late May/June 45 cm apart.	February to April	Fruits are used cooked or raw in salads (sweet peppers) and some will mature outdoors in warm sunny areas but are best under cloches or glass elsewhere. July to September.
Potatoes	TUBERS (seed potatoes: plant 15 cm deep)	Sprout earlies in shallow boxes in full light and frost-free place before planting out. Plant out from March onwards 25–30 cm apart in rows 60 cm apart for earlies, 90 cm for main crop.	February onwards	Tubers used cooked. Can be available all year starting with earlies in late June and storing maincrops in outdoor 'clamps' over winter. Large tubers can be divided but this is not recommended by many growers nowadays..
Pumpkins	SEED (sow 1.3 cm deep)	Sow singly in 8 cm pots under glass and plant out late May 1.5 m apart or sow 2–3 seeds outdoors at each station in May and thin to 1.	March/April	Use fruits as a vegetable marrow when young or grow to maturity and store. July onwards.
Radish	SEED (sow 0.5 cm deep	Sow in drills 20 cm apart (summer)	March to July	Use raw in salads sowing frequently for succession. Can be harvested all year with the right varieties and treatment.
Winter Types	SEED (sow 2.5 cm deep)	Sow in drills 23 cm apart and thin to 15 cm in the row. Lift to store in November.	July/August	
Runner Beans	SEED (sow 7.5 cm deep)	Sow in double rows 30–38 cm apart spacing plants 23–30 cm apart. Outdoors sow 2 seeds together and thin to 1 seedling later.	May/June	Tall climbers needing support. Pods are used cooked. July until September/October. Can be raised in small fibre pots under glass early May to plant out when spring frosts have gone.
Salsify	SEED (sow 2.5 cm deep)	Sow in rows 30–38 cm apart and thin to 15–20 cm in the row.	April/May	Roots used cooked and young leaves can be eaten in salads. Lift and store in November. November to April.
Scorzonera	SEED (sow 2.5 cm deep)	Sow in rows 30–38 cm apart and thin to 20–30 cm in the row.	April/May	Roots used cooked and leaves raw in salads. Lift as required December to April.
Shallots	SETS (bulblets)	Press bulbs into soil surface in rows 30 cm apart, spacing bulbs every 23 cm.	January to March	Mostly grown for pickling though of a milder flavour than onions when cooked. July onwards, lifting for storage in September.
Spinach	SEED (sow 2.5 cm deep)	Sow in rows 30 cm apart and thin to 23 cm in the row.	March to July	Grow true spinach in slight shade. Can be cropped all year with the right varieties grown for succession. Leaves used cooked. New Zealand type useful for hot dry sites.
New Zealand Spinach	SEED (soak 24 hours in clean water before sowing)	Sow under glass and prick out into small fibre pots. Plant out in May 60 cm apart in rows 90 cm apart.	March/April	
Swede or Rutabaga	SEED (sow 1.3 cm deep)	Sow in rows 45 cm apart and thin to 23–30 cm in the row.	May/June	Roots used cooked. Leave in ground and harvest as required or lift and store. November onwards.

PLANT NAME	METHOD OF PROPAGATION	TREATMENT REQUIRED	TIME OF YEAR PROPAGATED	CULTURAL NOTES
Sweetcorn (*Corn-on-the-Cob*)	SEED (sow 3.8 cm deep)	Sow under cloches 45 cm apart, planting 2–3 seeds at each station and later thin to 1.	April/May	Seedheads used cooked. Can be raised from March onwards under glass 1–2 seeds to a small pot. Plant out late May in blocks rather than rows. August to November.
Swiss Chard (*Rhubarb Chard*)	SEED (sow 2.5 cm deep)	Sow in rows 45 cm apart and thin to 23–30 cm in the row.	April to July	Leaves and leaf stalks used cooked. Crop best protected by cloches over winter. All year round.
Tomatoes	SEED (for growing on under glass)	Sow in a temperature of 16–18°C. Prick out and pot on as required. For cold house sow a little later.	January onwards	Fruit used cooked or raw. Grow under glass 45–60 cm apart or use outdoor varieties in a warm sheltered position outdoors or under cloches. June to November.
	SEED (for outdoors)	Sow under glass and harden off before planting out in June.	March/April	
Tobacco	SEED (surface sow)	Sow thinly in a temperature of 16°C. Harden off and plant out 90 cm apart in late May/early June.	February/early March	Leaves are gathered in late summer/autumn and used appropriately prepared for smoking. Detailed instructions needed for successful results.
Turnip	SEED (sow 1.3 cm deep)	Sow in rows 30 cm apart and thin to 15–20 cm in the row. Can be sown earlier under cloches.	April to August	Roots used cooked when about tennis ball size. Lift and store in October/November for winter storage. Can be available all the year round.

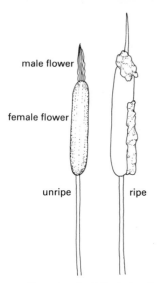

male flower

female flower

unripe

ripe

Flowerheads of *Typha* (reedmace)

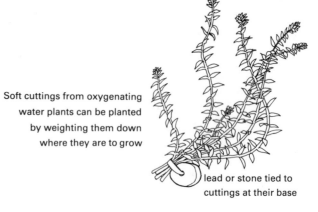

Soft cuttings from oxygenating water plants can be planted by weighting them down where they are to grow

lead or stone tied to cuttings at their base

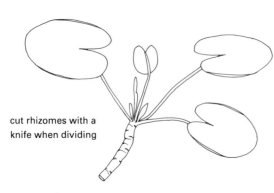

cut rhizomes with a knife when dividing

Water lilies (*Nymphaea*)

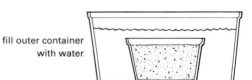

Fill inner pot with a compost of sandy soil mixed with a little charcoal and sow seed in the normal way

fill outer container with water

Sowing aquatic plant seeds

replant divisions or plant seedlings out in 'baskets'

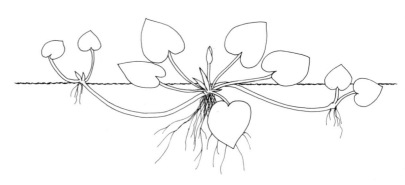

Hydrocharis morsus-ranae produces runners which can be separated from the parent plant

20 Water Garden Plants

A pool, whether ornamental or natural, adds yet another dimension to the garden and the plants that grow in and around them present no more cultural difficulty than their more terrestial-based relations.

Propagation in most cases is much the same too, the difference lying in the amount of moisture needed either for the germination of seed or the aftercare of young plants. Many marginals (shallow water plants) and bog plants are herbaceous perennials, and seedlings etc are best grown on in nursery areas to establish them before planting out. The same applies to the true aquatics for pool or aquarium. If a tank or something similar can be set aside for propagating and establishing the choicer specimens then so much the better. It need not be very large or very deep but remember to shade in very sunny weather as small areas of water soon heat up to a perhaps fatal degree.

Marginals and bog plants are planted out as for other garden subjects, but aquatics need to be 'placed' in position. The free-floating types are simply placed on the surface of the water but the rooting types have to be planted either in the special plastic mesh baskets, or by weighting with a stone or piece of lead and simply dropping into the pool in the required place where the new plant can take root on the bottom. Fill baskets or base and shelves of new pools with 10–15 cm of ordinary garden soil, preferably containing little fibrous material or soluble fertilisers. For bog gardens make up a mixture of equal parts by volume loam/coarse moss peat with a little slow release organic fertiliser.

Outdoors, pools should be sited in full sun and away from tree leaf-fall if possible, and unless stated the plants in the following pages are hardy and prefer this situation. Although this chapter contains the majority of true aquatics likely to be met with, there are many other marginal or moisture-loving plants distributed among genera mentioned elsewhere in the book and duly noted where there has been space.

PLANT NAME	METHOD OF PROPAGATION	TREATMENT REQUIRED	TIME OF YEAR PROPAGATED	CULTURAL NOTES
Alisma (Water Plantain)	DIVISION	Divide and replant in wet soils.	April to June or in autumn	Perennial flowering plant for marginal or bog conditions. Species from 15 to 60 cm in height. Will self-seed.
	SEED	Sow in fibrous soil with a little charcoal mixed in just below water surface.	When ripe or in spring	
Anubias	DIVISION (of rhizome)	Replant in sand/clay compost.	At any time	Low, slow-growing marsh and submerged aquatics for aquaria or terraria around 21°C.
Aponogeton (A. distachyus = Water Hawthorn)	DIVISION (of tuber)	Divide and replant according to species.	April to June	Aquatic flowering plants for water 5 to 60 cm deep depending on the species. Annuals and perennials, some such as A. distachyus and A. krauseanus being hardy, others tender.
	SEED	Sow in fine soil with charcoal added in shallow pans standing in water in a temperature of 13–16°C.	As soon as ripe	
Azolla (Fairy Moss; A. caroliniana)	DIVISION	Separate clumps and replace on water surface. Overwinter in shallow water over mud or sand in a frost-free greenhouse.	At any time when in active growth	Small floating ferns grown for foliage effect. Grow on shallow pools outdoors or in aquariums indoors. Rather tender.
Butomus (B. umbellatus = Flowering Rush)	DIVISION	Replant in shallow water. Can also be raised from seed sown in spring.	April to June	A flowering perennial up to 90 cm in height for pond and lake margins.
Cabomba (Fanwort)	TIP CUTTINGS	Plant 5 cm deep in aquarium substrate.	At any time when in active growth	C. carolinia is a submerged oxygenator for cold water aquaria. Others need heat.
Calla (C. palustris = Bog Arum)	DIVISION	Separate sections of creeping rootstock and replant in wet soils.	April to June	Flowering/berry-bearing perennials for marginal or bog conditions. Grows 15–30 cm in height.
Callitriche (Water Starwort)	SOFTWOOD CUTTINGS	Attach to weight and drop into pool or aquarium as required. Can also be divided.	In spring and early summer	Oxygenating aquatic perennials for ponds or aquaria of any depth.
Caltha (Marsh Marigold/Kingcup)	SEED	Sow in pots stood in shallow water in a cold frame.	March to May	Perennial flowering plants with species up 90 cm in height. Grow on lake or pool banks in sun or slight shade.
	DIVISION	Replant in moist soils around 30 cm apart.	April to June after flowering	
Cryptocoryne (Water Trumpet)	DIVISION (offsets: runners)	Separate runners and replant singly 2.5 cm deep in the substrate of pool or aquarium.	At any time	Many species of oxygenators and bog plants for heated pools and aquariums.
Decodon (syn. Nesaea)	SOFTWOOD CUTTINGS	Root in moist soil in a cold frame.	In spring	A flowering perennial up to 60 cm in height. Pondside plant.
Echinodorus (E. paniculatus = Amazon Sword)	DIVISION (offsets: runner)	Replant in wet soils or mud outdoors or aquarium species in the substrate. All can also be raised from seed.	April to June	E. ranunculoides is a hardy flowering perennial for marginal plantings. There are other tender species for heated aquaria.
Eichhornia (E. crassipes = Water Hyacinth)	DIVISION (of plant or stolons)	E. crassipes can be placed on water surface. Others plant in soil or baskets in water 30 cm deep. Place E. crassipes outdoors in June. Overwinter at 13–16°C.	May to August	Mostly tender flowering perennials with floating leaves needing temperatures around 16–24°C though E. crassipes can be grown outdoors in summer.
Eleocharis (E. acicularis = Hair Grass/Needle Spike-Rush)	DIVISION (offsets: runners)	Replant in moist gravelly soil or in shallow water.	In spring outdoors or at any time indoors	Indoor aquarium or outdoor pool plant either for bog or as submerged aquatic.

PLANT NAME	METHOD OF PROPAGATION	TREATMENT REQUIRED	TIME OF YEAR PROPAGATED	CULTURAL NOTES
Elodea (Anacharis/Canadian Pondweed)	DIVISION	Divide and replant by weighting and sinking in position desired.	April to June	Submerged oxygenating plants for aquaria or pools.
	STEM CUTTINGS	Cut stems and replant as above.	In spring	
Gillenia (G. trifoliata = Indian Physic)	DIVISION	Replant in moist soil. Can also be raised from seed or basal cuttings rooted in a propagator in spring.	March/April	Flowering perennials around 60 cm in height for moist peaty soils and shady positions.
Glyceria (G. aquatica, syn. Catabrosa aquatica)	DIVISION	Replant in shallow water.	April to June	Perennial grass for marginal plantings. Grows to 60 cm in height.
Gunnera (Prickly Rhubarb)	SEED	Sow in sand/peat in a temperature of 16–18°C.	March to May	Mostly massive flowering herbaceous perennials mostly grown for their large leaves. .G. magellanica is a mat-forming plant about 8 cm in height. All may need some winter protection in cold areas.
	DIVISION	Replant in moist soils in sheltered positions at pool or lakeside. Overwinter seedlings and small divisions in a cold frame.	April	
Hottonia (Water Violet)	DIVISION	Set divisions in pots or small baskets and sink where required to grow.	April to June	Aquatic flowering and oxygenating herbaceous perennial for shallow water in pond or aquarium.
Houttuynia	DIVISION	Replant 23–30 cm apart in wet boggy soil.	March/April	Flowering perennial up to 60 cm in height for pool margins or shallow water in slight shade.
	ROOT CUTTINGS	Establish in a cold frame in wet soil.	December/January	
Hydrocharis (H. morsus-ranae = Frogbit)	DIVISION (rooted runners)	Replace divisions on surface of water.	April to June	Very small floating flowering aquatic for outdoor pools.
Hydrocleys (Water Poppy)	SEED	Sow in pots of rich soil and sink just below water surface.	In spring	Aquatic flowering perennials with floating leaves. Best lifted and overwintered under frost-free glass in colder areas. Can also be grown from cuttings.
	DIVISION	Replant in pots or baskets and sink in 15 cm of water.	April to June	
Hygrophila (H. polysperma)	LEAF CUTTINGS	Remove leaf with stem and root in warm water.	At any time	Fast-growing bog or aquatic flowering plants for heated aquaria.
Lemna (Duckweed)	DIVISION	Separate small groups and replace on water surface.	At any time while in growth	Dwarf floating aquatics for ponds or aquaria.
Limnobium	DIVISION (offsets: runners)	Cut runners and replace separated plants on water surface.	At any time when in growth	Small floating aquatics with lily-like leaves for heated aquaria indoors.
Ludwigia	STEM CUTTINGS	Root in substrate in shallow water.	During summer	L. repens is a submerged and flowering aquatic for aquaria.
Lysichitum (L. americanum = Skunk Cabbage)	SEED	Sow in a cold frame or greenhouse. Keep compost very wet.	As soon as seed is ripe	Flowering perennials around 60 cm in height for marginal plantings in sun or slight shade.
	DIVISION	Replant in moist boggy soils 60 cm apart.	After flowering	
Lythrum (L. salicaria = Purple Loosestrife)	SEED	Sow outdoors or in a cold frame.	May/June	Flowering perennials around 90 cm in height for moist soils in sun or slight shade. Good in moist borders or at the pondside.
	DIVISION	Replant 45 cm apart.	March/April or October	
	BASAL CUTTINGS	Root 5–8 cm cuttings in peat/sand in cold frame or propagator.	March to May	

PLANT NAME	METHOD OF PROPAGATION	TREATMENT REQUIRED	TIME OF YEAR PROPAGATED	CULTURAL NOTES
Menyanthes (M. crista-galli = Cock's Comb; M. trifoliata = Bogbean)	DIVISION/CUTTINGS	Insert sections of creeping stem into mud.	April/May	Aquatic flowering perennials with floating leaves for shallow water in all situations.
Mimulus (Musk/Monkey Flower)	SEED	Sow shallowly in a temperature of 16–18°C.	February/March	Flowering herbaceous perennials, some sometimes grown as annuals for moist shady borders. Many species from 15 cm to 1.5 m in height. Grow well in bog gardens.
	SOFTWOOD CUTTINGS	Root in a sandy compost in a propagator.	April to June	
	DIVISION	Replant according to the species in moist soils in slight shade.	March/April	
Myriophyllum (Water Milfoil)	SOFTWOOD CUTTINGS	Root in pans of soil submerged in shallow water. Can also be divided.	Spring and summer	Oxygenating aquatics and marginals for ponds and aquariums. Tender and hardy species.
Nelumbo (syn. Nelumbium; N. nucifera = East Indian Lotus)	SEED (chip or file)	Sow singly in sandy soil in 8 cm pots placed 5–8 cm below water surface at 18–24°C.	April	Large aquatic flowering perennials needing water at 16–18°C and a frost-free environment in winter. Can be grown in pools outdoors in mild areas with adequate winter protection.
	DIVISION (of rhizome)	Replant in rich soil 2.5–5 cm deep in shallow water or deeper in all respects outdoors.	April to June	
Nuphar (N. advena = Common Spatterdock; N. lutea = Yellow Water Lily)	DIVISION (of corm-like tuber)	Replant divisions in rich soil in baskets and sink in position.	April to June	Water-lily-like flowering aquatics for sun or slight shade in 15–30 cm of water, though some will grow in deeper.
	SEED	Sow in a cold frame in loam with a little charcoal and keep wet.	As soon as seed is ripe	
Nymphaea (Water Lily)	DIVISION (of tuber/rhizome)	Cut with a knife or detach offsets. Replant in rich loam in baskets.	April to June	Flowering aquatic perennials with floating leaves. Many species and varieties for water 10 to 60 cm deep according to the type. The tender types require 10–16°C in winter.
	SEED	Sow in loam with a little charcoal. Keep wet. Cold frame for hardy types, 18–24°C for tender.	As soon as ripe	
Nymphoides (Limnanthemum: L. nymphoides, syn. N. peltata = Fringed Water Lily)	SEED	Sow in wet soil, hardy types in a cold frame tender types at 16–18°C.	April/May	Flowering aquatics with small floating lily-like leaves. Hardy and tender species N. peltata being one of hardy ones. Tender types need 7°C in winter.
	DIVISION	Replant in baskets and sink in position.	April to June	
Orontium (O. aquaticum = Golden Club)	SEED	Sow in a cold frame and keep compost wet.	As soon as seed is ripe	Flowering perennial for ponds or slow-moving water up to 45 cm in depth. Plant grows above water when shallow, floats in deep.
	DIVISION	Replant in deep loamy soils.	April to June	
Peltandra (Arrow Arum)	DIVISION	Replant in groups in shallow water.	April to June	Perennial flower and berry-bearing plants for marginal plantings.
Peltiphyllum (Umbrella Plant)	SEED	Sow in a cold frame keeping compost moist.	March to May	Flowering perennial up to 1.2 m in height. Has large umbrella-like leaves. Ideal for the waterside.
	DIVISION	Replant in moist soils.	March/April	
Penthorum (P. sedoides = Joseph's Coat/Virginian Stonecrop)	DIVISION	Replant in moist soils or shallow water. Can also be grown from cuttings.	April to June	P. sedoides is a hardy flowering perennial 45–60 cm in height for marginal or waterside plantings.
Pistia (Water Lettuce)	DIVISION (runners)	Separate and replace on water surface.	At any time when in active growth	Floating aquatics for heated pools and aquaria indoors.

PLANT NAME	METHOD OF PROPAGATION	TREATMENT REQUIRED	TIME OF YEAR PROPAGATED	CULTURAL NOTES
Pontederia (*P. cordata* = Pickerel Weed)	DIVISION	Replant in rich soil in shallow water 23–30 cm apart.	April to June	Marginal/aquatic perennials grown for foliage and flowers. *P. cordata* is hardy but some species are tender.
Potamogeton (Pondweed)	SOFTWOOD CUTTINGS	Tie to weights and sink in position where required.	April to June	Submerged aquatics for oxygenating pools, lakes, slow-moving water and aquaria.
	DIVISION	Replant divisions as above.	April to June	
Ranunculus (*R. aquatilis* = Water Crowfoot)	DIVISION	Replant *R. lingua* in moist soils, *R. aquatilis* by weighting and sinking in position. Position 23 cm apart.	April to June	*R. aquatilis* is a floating and submerged oxygenator for pools and slow-moving water. *R. lingua* is a bog or marginal plant.
Sagittaria (Arrowhead)	SEED	Sow in a constantly wet compost in slight heat.	March/April	Aquatic flowering/foliage plants with hardy and tender species all around a metre in height. Grow tender species in pots and house in frost-free position in winter.
	DIVISION (offsets)	Replant around 30 cm apart in 8–15 cm of water in lakes and ponds.	April to June	
Salvinia	DIVISION	Divide groups and move to fresh sites on water surface. No soil necessary.	At any time while in growth	Small tender aquatic ferns, surface floating and rooting. Grow in warm greenhouse or indoor aquarium. 13°C minimum in winter.
Saururus (*S. chinensis* = Lizard's Tail)	DIVISION	Replant in heavy loam in 8–15 cm of water at pond margins.	April to June	Flowering aquatic perennials with heart shaped leaves. Up to 60 cm in height.
Scirpus (Bulrush: *S. cernuus*, syn. *Isolepis gracilis* = Clubrush)	DIVISION (of rhizome)	Replant in boggy soils or shallow water at the margins of pools lakes and streams.	April to June	Marsh or aquatic plants with hardy and tender species, these latter requiring a minimum of 7°C in winter.
Stratiotes (Water Soldier)	DIVISION (offsets: stolons)	Separate when new growth is growing strongly and replace on water surface where required.	Spring and summer	*S. aloides* is a flowering aquatic with floating leaves. Young plants separate naturally in autumn, sink and overwinter on pool bottom.
Trapa (*T. natans* = Water Chestnut; spiny fruits are edible)	SEED	Sow in pans of loam in water where plants are to grow. Heat to around 18–21°C until germination occurs then grow on in unheated water.	April/May	Floating aquatics, most for tanks and pools in a cool greenhouse 7°C in winter though *T. verbanensis* is hardy.
Typha (Reedmace; *T. minima* = Dwarf Reedmace)	DIVISION (of rhizome)	Replant in moist soils 15 cm apart (*T. minima*).	April to June	Mostly tall aquatic perennials for pond margins in water up to 15 cm deep. *T. minima* best for gardens.
	SEED	Sow in sand/peat kept wet in a cold frame.	April/May	
Utricularia (Bladderwort)	DIVISION	Replant divisions by dropping them in water where they are to grow.	April to June	Submerged flowering aquatics for ponds up to 60 cm in depth. Some species are carnivorous.
Vallisneria (*V. spiralis* = Eel Grass)	DIVISION (offsets: runners)	Replant in rich soil in baskets etc. at the bottom of pools, aquaria or tanks.	April to June	Submerged aquatic oxygenators with grass-like leaves for pools or aquaria indoors or in mild areas only outdoors. Require a 4°C minimum.

Metric Conversion Table

Length

13 mm = $\frac{1}{2}$ in	23 cm = 9 in
2.5 cm = 1 in	30 cm = 1 ft
5 cm = 2 in	45 cm = $1\frac{1}{2}$ ft
10 cm = 4 in	60 cm = 2 ft
15 cm = 6 in	90 cm = 3 ft

Weight

28 g = 1 oz	1 kg = 2 lb 3 oz

Temperature

5°C = 41°F	20°C = 68°F
10°C = 50°F	25°C = 77°F
15°C = 59°F	30°C = 86°F

Index of Common Names

Index of Common Names

Index of Common Names

Index of Scientific Names

Abelia 168
Abeliophyllum 168
Abies 168
Abutilon 152
Acacia 152
Acaena 12, 52
Acalypha 152
Acanthocalcium 98
Acanthocereus 97
Acantholimon 52
Acanthus 11, 134
Acer 168
Achillea 134
Achimenes 76, 78
Acidanthera 78
Aconitum 134
Acorus 126
Acroclinium 68
Acrostichum 108
Actaea 134
Actinella 52
Actinidia 168
Actinomeris 134
Actinopteris 108
Adenanthera 152
Adenium 96
Adenophora 134
Adiantum 108
Adonis 134
Adromischus 96
Aechmea 152
Aeonium 96
Aeschynanthus 152
Aesculus 168
Aethionema 52
Agapanthus 78
Agastache 134
Agàve 96
Ageratum 68
Aglaomopha 108
Aglaonema 152
Agrostemma 68
Aichrýson 96
Ailanthus 168
Ajuga 134
Akebia 168
Albizzia 152
Albuca 78
Alchemilla 134
Alisma 202
Allium 16, 25, 76, 78
Alnus 168
Alocasia 152
Aloe 96
Alonsoa 68
Alpinia 152
Alsophila 108
Alstroemeria 78
Althaea 134
Alyssum 52, 68
Amaranthus 68

Amaryllis 78, 86
Amasonia 152
Amelanchier 168
Ammobium 68
Ammocharis 78
Amorpha 168
Amsonia 135
Anacampseros 96
Anacharis 203
Anagallis 52
Ananas 152
Anaphalis 135
Anchusa 135
Ancistrocaçtus 96
Andromeda 170, 183
Androsace 52
Andryala 52
Aneimia 108
Anemone 15, 78, 135
Anemonopsis 135
Angiopteris 108
Antennaria 52
Anthemis 135
Anthericum 79
Antholyza 79
Anthurium 152
Anthyllis 53
Antigonon 153
Antirrhinum 68
Anubias 202
Aphelandra 153
Aponogeton 202
Aporocactus 96
Aquilegia 15, 135
Arabis 53
Arachis 153
Aralia 168
  sieboldii 157
Araucaria araucana 169
  excelsa 153
Araujia 153
Arbutus 169
Archontophoenix 153
Arctotis 68
Ardisia 153
Arenaria 53
Argyroderma 96
Ariocarpus 96
Arisaema 79
Arisarum 53
Aristea 153
Aristolochia 153
Armeria 12, 50, 53
Arnebia 135
Artemisia 135
Arthropodium 79
Arum 79
Aruncus 135
Arundinaria 169
Asclepias 135, 153
Asparagus 153

Asperella 141
Asperula 53
Asphodelus 79
Aspidistra 153
Asplenium 108
Aster 136
Astilbe 136
Astragalus 136
Astrantia 136
Astroloba 96
Astrophytum 96
Athyrium 108
Atriplex 68
  halimus 169
Aubretia (Aubrieta) 53
Aucuba 169
Avena 141
Azalea 185
  indica 153
Azara 169
Azolla 202

Babiana 79
Bahia 56
Baptisia 136
Bartonia 68
Bauhinia 153
Beaufortia 153
Beaumontia 154
Begonia 9, 10, 68, 79
Belamcanda 79
Bellis 136
Beloperone 154
Berberidopsis 169
Berberis 166, 169
Bergenia 136
Berkheya 136
Bessera 80
Betula 169
Billbergia 154
Blechnum 108
Bletilla 60
Bloomeria 80
Blossfeldia 96
Bobartia 80
Boltonia 136
Bomarea 80
Boronia 154
Borzicactus 96
Bougainvillea 154
Bouteloua 141
Bowiea 80
Boykinia 136
Brachycombe 68
Brainea 108
Brassaia 163
Bravoa 80
Brodiaea 80
Browallia 154
Brunfelsia 154
Brunnera 136

Brunsvigia 80
Bryophyllum 94, 100
Buddleia 169
Bulbine 80
Bulbinella 80
Bulbocodium 80
Buphthalmum 136
Bupleurum 169
Butomus 202
Buxus 169

Cabomba 202
Cacalia 70
Cactus 96
Caesalpinia 154
Caladium 154
Calamintha 54
Calandrinia 68
Calathea 154
Calceolaria 154
  integrifolia 69
Calendula 69
Caliphruria 80
Calla 202
Callicarpa 169
Callipsyche 81
Callirhoe 137
Callistemon 154
Callistephus 69
Callitriche 202
Calluna 170
Calocedrus 179
Calochortus 81
Calostemma 81
Caltha 202
Calycanthus 170
Calypso 60
Camellia 170
Cammassia 81
Campanula 137
  medium 69
Campsis 12, 170
Camptosorus 108
Canna 154
Capsicum 154
Caragana 170
Caralluma 97
Cardamine 137
Cardiocrinum 81
Cardiospermum 69
Carduncellus 137
Carex 154
Carica 154
Carlina 137
Carnegiea 97
Carpenteria 170
Carpinus 170
Carpobrotus 97
Carya 170
Caryopteris 170
Cassia 155

213

Index of Scientific Names

Index of Scientific Names